THE
TRUTH ABOUT
CONTAGION

THE
TRUTH ABOUT
CONTAGION

THE
TRUTH ABOUT
CONTAGION

Exploring Theories of
How Disease Spreads

THOMAS S. COWAN, MD, and
SALLY FALLON MORELL

Skyhorse Publishing

10 9 8 7 6 5 4

Library of Congress Cataloging-in-Publication Data is available on file.

Print ISBN: 978-1-5107-6882-6
Ebook ISBN: 978-1-5107-6791-1

Printed in the United States of America

Disclaimer

The information contained herein should NOT be used as a substitute for the advice of an appropriately qualified and licensed physician or other health care provider. The information provided here is for informational purposes only. Although we attempt to provide accurate and up-to-date information, no guarantee is made to that effect. In the event you use any of the information in this book for yourself, the authors and the publisher assume no responsibility for your actions.

CONTENTS

PREFACE
by Sally Fallon Morell

Since the dawn of the human race, medicine men and physicians have wondered about the cause of disease, especially what we call "contagions." Numerous people become ill with similar symptoms, all at the same time. Does humankind suffer these outbreaks at the hands of an angry god or evil spirit? A disturbance in the atmosphere? A miasma? Do we catch the illness from others or from some outside influence?

With the invention of the microscope in 1670 and the discovery of bacteria, doctors had a new candidate to blame: tiny one-celled organisms that humans could pass from one to another through contact and exhalation. But the germ theory of disease did not take hold until two hundred years later with celebrity scientist Louis Pasteur and soon became the explanation for most illness.

Recognition of nutritional deficiencies as a cause of diseases like scurvy, pellagra, and beriberi took decades because the germ theory became the explanation for everything that ails the human being. As Robert R. Williams, one of the discoverers of thiamine (vitamin B1) lamented, "all young physicians were so imbued with the idea of infection as the cause of disease that it presently came to be accepted as almost axiomatic that disease could have no other cause [other than microbes]. The preoccupation of physicians with infection as a cause of disease was doubtless responsible for many digressions from attention to food as the causal factor of beriberi."[1]

During the Spanish flu pandemic of 1918, the deadliest example of a contagion in recent history, doctors struggled to explain the world-wide reach of the illness. It sickened an estimated five hundred million people—about one-third of the planet's population—and killed between twenty to fifty million people. It seemed to appear spontaneously in different parts of the world, striking the young and healthy, including many American servicemen. Some communities shut down schools, businesses, and theaters; people were ordered to wear masks and refrain from shaking hands, to stop the contagion.

But was it contagious? Health officials in those days believed that the cause of the Spanish flu was a microorganism called *Pfeiffer's bacillus,* and they were interested in the question of how the organism could spread so quickly. To answer that question, doctors from the US Public Health Service tried to infect one hundred healthy volunteers between the ages of eighteen and twenty-five by collecting mucous secretions from the noses, throats, and upper respiratory tracts of those who were sick.[2] They transferred these secretions to the noses, mouths, and lungs of the volunteers, but not one of them succumbed; blood of sick donors was injected into the blood of the volunteers, but they remained stubbornly healthy; finally they instructed those afflicted to breathe and cough over the healthy volunteers, but the results were the same: the Spanish flu was not contagious, and physicians could attach no blame to the accused bacterium.

Pasteur believed that the healthy human body was sterile and became sick only when invaded by bacteria—a view that dominated the practice of medicine for over a century. In recent years we have witnessed a complete reversal of the reigning medical paradigm—that bacteria attack us and make us sick. We have learned that the digestive tract of a healthy person contains up to six pounds of bacteria, which play many beneficial roles—they protect us against toxins, support the immune system, help digest our food, create vitamins, and even produce "feel good" chemicals. Bacteria that coat the skin and line the vaginal tract play equally protective roles. These discoveries call into question many current medical practices—from antibiotics to hand washing. Indeed, researchers have become increasingly frustrated in their attempts to prove that bacteria make us sick, except as coactors in extremely unnatural conditions.

Enter viruses: Louis Pasteur did not find a bacterium that could cause rabies and speculated about a pathogen too small for detection by microscopes. The first images of these tiny particles—about one-thousandth the size of a cell—were obtained upon the invention of the electron microscope in 1931. These viruses—from the Latin *virus* for "toxin"—were immediately assumed to be dangerous "infectious agents." A virus is not a living organism that can reproduce on its own, but a collection of pro-teins and snippets of DNA or RNA enclosed in a membrane. Since they are seen in and around living cells, researchers assumed that viruses rep-licate only inside the living cells of an organism. The belief is that these ubiquitous viruses "can infect all types of life forms, from animals and plants to microorganisms, including bacteria and archaea."[3]

Difficult to separate and purify, viruses are a convenient scapegoat for diseases that don't fit the bacterial model. Colds, flu, and pneumo-nia, once considered exclusively bacterial diseases, are now often blamed on a virus. Is it possible that scientists will one day discover that these particles, like the once-maligned bacteria, play a beneficial role? Indeed, scientists have already done just that, but old ideas, especially ideas that promise profits from drugs and vaccines—the "one bug, one drug" men-tality—die hard.

Today, the premise that coronavirus is contagious and can cause dis-ease has provided the justification for putting entire nations on lockdown, destroying the global economy and throwing hundreds of thousands out of work. But is it contagious? Can one person give coronavirus to others and make them sick? Or is something else, some outside influence, caus-ing illness in the vulnerable?

These questions are bound to make public health officials uncom-fortable—even angry—because the whole thrust of modern medicine derives from the premise that microorganisms—transmittable microor-ganisms—cause disease. From antibiotics to vaccines, from face masks to social distancing, most people submit willingly to such measures in order to protect themselves and others. To question the underlying principle of contagion is to question the foundation of medical care.

I am delighted to join my colleague Tom Cowan in creating this exposé of the modern medical myth—that microorganisms cause disease and that these diseases can be spread from one person to another through coughs, sneezes, kisses, and hugs. Like Tom, I am no stranger to contro-versial views. In my book *Nourishing Traditions*, first published in 1996, I proposed the heretical idea that cholesterol and saturated animal fats are not villains, but essential components of the diet, necessary for normal growth, mental and physical well-being, and the prevention of disease.

In *Nourishing Traditions* and in other writings, I presented the rad-ical notion that pasteurization—collateral damage of the germ the-ory—destroys the goodness in milk and that raw whole milk is both safe and therapeutic, especially important for growing children. It is the most obvious substitute for breast milk when mothers are having trouble nursing their babies, a proposition that makes health officials squirm. In subsequent publications I have argued the dissenting view that it is a nutrient-dense diet and not the administration of vaccines that best protects our children from illness. Over the years these views have found increasing support with both laypeople and health professionals.

Error has consequences. The result of the notion that our diets should be devoid of animal fats, that children should grow up on processed skim milk, and that it's fine to vaccinate them dozens of times before the age of five has resulted in immense suffering in our children, an epidemic of chronic illness in adults, and a serious decline in the quality of our food supply. There are economic consequences as well, including the devastation of rural life as small farms, especially dairy farms prohibited from selling their milk directly to customers, give in against the price pressures of "Big Ag" (Big Agriculture/corporate farming), and parents of children with chronic illness (estimated to be as high as one child in six[4]) struggle with the costs of caring for them.

What are the possible consequences of the premise that microorganisms, especially viruses, cause disease? The "coronavirus pandemic" gives us many clues: forced vaccinations, microchipping, prescribed social distancing, lockdown, mandatory masks, and negation of our right to assemble and practice our religion whenever an illness appears that can be media hyped into a public health emergency.

Until we base our public policies on the truth, the situation will only get worse. The truth is that contagion is a myth; we need to look elsewhere for the causes of disease. Only when we do so will we create a world of freedom, prosperity, and good health.

—Sally Fallon Morell
July 2020

INTRODUCTION
by Thomas S. Cowan, MD

I am no stranger to controversial views, particularly controversial positions in the field of medicine. In my latest series of three books, I have denounced several sacred icons that form the basis of our attitudes toward disease and its treatment.

In *Human Heart, Cosmic Heart*, I clearly demonstrated that the heart is not a pump and that blocked arteries are not the predominant cause of heart attacks.

Then, in *Vaccines, Autoimmunity, and the Changing Nature of Childhood Illness*, I proposed the theory that acute illness is not caused by an infection that attacks us from the outside but rather represents a cleansing of our watery, cellular gels. A corollary to this position is that any intervention that interferes with this cleansing response, in particular vaccines, is bound to create untold harm that manifests in skyrocketing rates of chronic disease.

In what I thought would be my third and final book, *Cancer and the New Biology of Water*, I show why the "war on cancer" is an utter failure. I argue that the modern chemotherapeutic approach to cancer is useless and that an entirely new way of looking at this problem must emerge. I postulated that this new way of looking at medicine and biology must put the question of "what actually causes disease" squarely in the forefront of our thinking.

I thought I was done with writing controversial books (at least about medicine) and that I could turn my attention to finishing out my career as a practicing physician; spending more time in the garden; and creating a healing place for myself, my friends, and my family. I knew I would continue doing occasional interviews and maybe some online classes or mentoring. I would still talk about the nature of water and the increasing pollution of our earth; but I also hoped that interest in my work would wane and simply become part of the general consciousness, a new way of thinking that would change our attitude toward disease and rehumanize the practice of medicine. I did have a nagging thought—which had been there for years—that I needed to delve into the HIV/AIDS affair, but I was content to let that be—it was more like an itch that only occasionally begged to be scratched.

Not long ago I had lunch with a homeopathic physician, and we were joking about our respective long careers in medicine, and how much things have changed over the years. For some reason, the conversation turned to immunology, and we asked each other what we remembered learning in medical school about immunology—that was back in the early 1980s. We both jokingly concluded that the only thing we remember was being taught that if you wanted to know whether a patient was immune to a particular viral disease, you could test antibody levels. If the antibodies were high, that meant they were immune.

Just as people remember for the rest of their lives the moment they heard that JFK was shot, or about the World Trade Center towers coming down on September 11, I have a vivid memory of hearing the announcement by Robert Gallo in 1984 that they had found the cause of AIDS. It was caused by a virus called HIV, and the reason they knew it caused AIDS is that they found elevated antibody levels in some (not all) AIDS patients. I remember turning to a fellow medical student at the time and saying, "Hey, who changed the rules?" In other words, after having spent the previous four years learning that people with antibodies to a virus were immune to that particular virus, we were now being told—with no explanation whatsoever—that antibodies meant that the virus was actually causing the disease!

I didn't buy it then, and I don't buy it now. For more than thirty-five years, I have read countless articles, books, papers, and documents about the lack of connection between HIV and AIDS. This naturally led me to investigate the connection between "viruses" and other diseases, and what I discovered was shocking, to say the least. That is the background of my now-famous ten-minute video about the cause of the coronavirus "pandemic."

Even though I have been aware for decades that the virus king is naked, I was hoping that others would take up the challenge to relay this information to the general public. But a ten-minute video thrust me onto the stage. It happened like this: in early 2020, I received an invitation to speak at a health conference in Arizona. I knew almost nothing about the group that invited me, but they gave me a first-class airplane ticket, so I agreed. I wasn't clear on what topic they wanted me to speak, but since I never speak with slides or notes, I figured I would improvise, as usual. Interestingly, a few times in the weeks leading up to this event my wife asked me where I was going, to whom I was speaking, and what the subject was. I just shrugged and said they seemed like nice, earnest people.

A few weeks earlier, the whole "coronavirus" event started to dominate the news. At first, I didn't think much of it, figuring that this was just another in a long line of viral scares—remember SARS, MERS, avian flu, Ebola, swine flu, and Zika? These were going to kill us all, but then just faded away.

But with "coronavirus," things started to intensify, particularly the dramatic, draconian responses by the authorities. Still, I didn't think much about it, although I did wonder whether the illnesses were the initial consequences of the planned 5G rollout—or perhaps a cover-up for the rollout. I thought about skipping the conference in Arizona, mostly because I was afraid of being quarantined there and not allowed to return home. I decided I was being paranoid and that I might as well honor my agreement to speak.

When I arrived at the conference, I discovered that there were only twenty or thirty attendees. The three other speakers had all canceled or decided to do their talks via Skype or Zoom. I was scheduled to do one talk each day of the two-day conference. The first day's talk was on acute illness and vaccines (my usual stump speech on that subject), with a talk on heart disease on the second day.

That night we started to hear more about quarantines and grounded planes. Given the sparse attendance, I spent part of that first night online to see whether I could catch an earlier flight home and just skip my second talk. I slept fitfully, worried about whether I should catch the 7 a.m. flight instead of my scheduled 1 p.m. flight. I decided that was crazy, and as long as I was there, I would do my talk on the heart and maybe end with a few comments about "viruses" and the current situation.

To say I didn't know I was being taped is not accurate, as I obviously wore a microphone and a guy in the back of the room seemed to be filming me, at least some of the time. But in my mind, I was clearly speaking to that group of twenty or thirty people. At the end of the talk, I made a

few off-the-cuff remarks about why viruses do not cause illness. I said my
piece and left for the airport. I was one of ten or so people on the plane,
and I made it safely home, very glad to be there.

A few days later, I got an email from Josh Coleman, the guy who
filmed the video, saying he had posted my remarks on viruses somewhere
online, and it was getting a huge response.[1] I thought that this might be
interesting but not much more. The rest, as they say, is history. I have
no idea how widely circulated that ten-minute video has become or how
many people have seen it—Josh tells me that it has had more than one
million views. I only knew that I needed to speak more about this subject, even if only to clarify what I had said at the conference.

Interest in my comments came from people all over the world.
Overnight I had become the point person for an alternative view of
viruses, the germ theory, the current health situation, and a lot more.
This led to a few podcast interviews, including one with Sayer Ji on
GreenMedInfo.com, and my own webinars.

Of course, I was criticized and even received some shocking threats,
but I have also received support in ways I could never have imagined.
I meant no harm to anyone. I am one man with a certain perspective,
hopefully correct in some things—and, if incorrect in other things, I
ask my readers only to understand that any errors come out of a place of
seeking the truth and my ability to understand the situation.

Two things press me forward. The first is to make it possible for
all of us to live in a world where everyone can speak their minds and
hearts freely without fear of recrimination or abuse. What could possibly be wrong with having an open and honest debate about the nature
and cause of illness and disease? This is a complex question, and no one
person or group has all the answers. But isn't that what real science, as
opposed to scientism, is supposed to be about?

Second, I am concerned that if my understanding of the current situation is even close to correct—an understanding for which we intend
to make a clear and convincing case in these pages—then humanity is at
a crossroads right now. There will be profound, even unimaginable consequences for all life on earth if we fail to heed the messages that emerge
from the current situation. My contention is that if we fail to understand
the true causes of the "coronavirus pandemic," we will go down a bitter
path from which there will be no turning back. That is what is driving
me to write this book.

I am happy to be writing this book with fellow iconoclast Sally Fallon Morell. Sally and I have been friends, collaborators (this is our third book together), and (I dare say) spiritual partners for over two decades. With a small contribution from me, Sally founded the Weston A. Price Foundation in 1999, perhaps the single best resource available for bringing truth in food, medicine, and farming to a world starving for that truth.

I sincerely wish this to be the last book Sally and I work on together. We have enjoyed collaborating, but I expect that the current "pandemic" we are living through will be a profound turning point in the history of humanity. It is my hope that out of this event, a new way of life will emerge in a world free of poisoned food, poisoned water, and the poisonous and false germ theory.

In this world, I envision no need for Sally and myself to write books. People will just know how to live; they will know that to poison their food, water, air, and the electric sheath of the earth is something only madmen can contemplate. We both look forward to the day when we can forget about warning people about this or that and spend more time growing and cooking food and sharing it in joy and laughter with our families, friends, and neighbors. No more books; after this, dear friends, you will know all you need to know.

Buckle up, folks, we are in for the ride of our lives.

—Thomas S. Cowan, MD
July 2020

PART 1

EXPOSING THE GERM THEORY

CHAPTER 1

CONTAGION

Let's get right to the nitty-gritty of this issue: contagion. How do we know whether any set of symptoms has an infectious cause? As we can all imagine, determining the cause of a disease in general, or a set of symptoms in any particular person, can be a complex and difficult task. Obviously, there are many factors to be considered for any one person at any one time in his or her life. Are the symptoms a result of genetics, poisoning, bad diet and nutrient deficiencies, stress, EMFs, negative emotions, placebo or nocebo effects—or infection from another person by a bacteria or virus?

In finding our way through this morass, we need well-defined rules to determine how to prove causation—and these rules should be clear, simple, and correct. We do have such rules, but scientists have ignored them for years. Unfortunately, failure to follow these guidelines threatens to destroy the fabric of society.

Imagine that an inventor calls you up and says he has invented a new ping-pong ball that is able to knock down brick walls and therefore make the process of demolition much easier and safer for builders and carpenters. Sounds interesting, although it is hard to imagine how a ping-pong ball could do such a thing. You ask the inventor to show you how he has determined that the new ping-pong balls are able to destroy brick walls. His company sends you a video. The video shows them putting a ping-pong ball in a bucket of rocks and ice cubes. They then take the bucket and fling it at a small brick wall. The wall goes down—"there's the proof," they say.

Wait a minute! How do we know it was the ping-pong ball that knocked the wall down and not the rocks and ice cubes that were also in the bucket?

"Good question," the inventor replies and then sends you a video showing an animated or virtual ping-pong ball destroying a virtual brick wall. He lets you know that the ball and the wall are exact renditions of the actual ball and brick. Still, something doesn't seem right; after all, it's fairly easy to create a computer image or video that shows such an occurrence, yet we would all agree it has nothing to do with what might happen with the actual ball and wall.

The inventor is getting exasperated with all your questions, but since you are a potential investor and he is interested in having your financial support, he persists. He then sends you a detailed analysis of what makes his ping-pong ball special. It has special protrusions on the outside of the ball that "grab onto and destroy the integrity of the cement holding the bricks together." Also, they build a lightweight internal system in the ping-pong ball that, according to the inventor, leverages the power of the ball, making it hundreds of times more powerful than the usual ping-pong ball. This, he says, is absolute proof that the new ball can whack down walls.

At this point, you are ready to hang up on this lunatic, but then he pulls the final trump card. He sends you videos of five esteemed researchers in the new field of ping-pong ball demolition. They, of course, have been funded entirely by the Ping-Pong Ball Demolition Council and have attained prestigious positions in the field. They each separately give testimony about the interesting qualities of this new ping-pong ball. They admit that more research is needed, but they have "presumptive" evidence that the claims of improved efficiency are correct and that a cautious investment is warranted. At that point, you do hang up the phone and check outside to see whether you've been dropped into Alice's Wonderland, and whether you have just been talking to the Mad Hatter.

Now if this ping-pong ball can really knock down brick walls, the obvious thing to do is to take the ping-pong ball, throw it at the wall, and record what happens—then have multiple other non-invested people do the same to make sure the company didn't put lead in the ball and throw it at a wall made of paper bricks. We could call this the Ultimate Ping-Pong Ball Test (UPPBT).

As bizarre and crazy as it sounds, this lack of evidence—that a microorganism called coronavirus pulls down the wall of your immune system, invades your cells, and starts replicating in them—is exactly what has happened with the "coronavirus" pandemic. No one has bothered to see what happens if you do the UPPBT, throwing the ball against the wall—and if you even suggest that we should do this, the trolls emerge from the shadows to call you a crazy person spreading "fake news."

Most people would agree with the requirement of proving that the ping-pong ball can destroy the brick wall; it's not something any of us would consider negotiable. And most people would agree that seeing a real brick wall demolished by a ping-pong ball constitutes proof. In other words, sane, rational human beings would accept the above UPPBT as true and relevant.

Heinrich Hermann Robert Koch (1843–1910) is considered one of the founders of modern bacteriology; he created and improved laboratory technologies for isolating bacteria and also developed techniques for photographing bacteria. His research led to the creation of Koch's postulates, a kind of UPPBT for disease, which consist of four principles linking specific microorganisms to specific diseases. Koch's postulates are as follows:

1. The microorganism must be found in abundance in all organisms suffering from the disease but not found in healthy organisms.
2. The microorganism must be isolated from a diseased organism and grown in a pure culture.
3. The cultured microorganism should cause disease when introduced into a healthy organism.
4. The microorganism must be re-isolated from the now diseased experimental host which received the inoculation of the microorganisms and identified as identical to the original specific causative agent.

If all four conditions are met, you have proven the infectious cause for a specific set of symptoms. This is the *only* way to prove causation. Interestingly, even Koch could not find proof of contagion using his postulates. He abandoned the requirement of the first postulate when he discovered carriers of cholera and typhoid fever who did not get sick.[1] In fact, bacteriologists and virologists today believe that Koch's sensible and logical postulates "have been recognized as largely obsolete by epidemiologists since the 1950s."[2]

Koch's postulates are for bacteria, not for viruses, which are about one thousand times smaller. In the late nineteenth century, the first evidence for the existence of these tiny particles came from experiments with filters that had pores small enough to retain bacteria and let other particles through.

In 1937, Thomas Rivers modified Koch's postulates in order to determine the infectious nature of viruses. Rivers' postulates are as follows:

1. The virus can be isolated from diseased hosts.
2. The virus can be cultivated in host cells.
3. Proof of filterability—the virus can be filtered from a medium that also contains bacteria.
4. The filtered virus will produce a comparable disease when the cultivated virus is used to infect experimental animals.
5. The virus can be re-isolated from the infected experimental animal.
6. A specific immune response to the virus can be detected.

Please note that Rivers drops Koch's first postulate—that's because many people suffering from "viral" illness do not harbor the offending microorganism. Even with Koch's first postulate missing, researchers have not been able to prove that a specific virus causes a specific disease using Rivers' postulates; one study claims that Rivers' postulates have been met for SARS, said to be a viral disease, but careful examination of this paper demonstrates that none of the postulates have been satisfied.[3]

Again, this book's central claim is that no disease attributed to bacteria or viruses has met all of Koch's postulates or all of Rivers' criteria. This is not because the postulates are incorrect or obsolete (in fact, they are entirely logical) but rather because bacteria and viruses don't cause disease, at least not in any way that we currently understand.

How did this state of error come about, especially concerning "infections" with bacteria and viruses? It goes back a long time—even to philosophies espoused in ancient Greece. Several philosophers and medics promoted this theory during the Renaissance,[4] but in modern times this masquerade became the explanation for most disease with that great fraud and plagiarist, Louis Pasteur, father of the germ theory.

Imagine a case in which some people who drink the milk from a certain cow develop profuse, bloody diarrhea. Your job is to find the cause of the problem. You wonder whether there is a transmissible agent in the milk that is being consumed by the unfortunate people, which makes them ill. This seems perfectly reasonable thus far. You then examine the milk under the newly invented microscope apparatus and find a bacterium in the milk; you can tell by its appearance that it is different from the usual bacteria that are found in all milk. You carefully examine the milk, discover that most if not all of the people with bloody diarrhea in fact did drink this milk. You then examine the milk consumed by people who didn't develop diarrhea and find that none of the milk samples contain this particular bacterium. You name the bacteria "listeria" after a fellow scientist. Then, to wrap up the case, you purify the bacteria, so that nothing else from the milk remains. You give this purified bacterial culture to a person who then develops bloody diarrhea; the clincher is that you then find this same bacteria in their stool. Case closed; infection proven.

Pasteur did this type of experiment for forty years. He found sick people, claimed to have isolated a bacterium, gave the pure culture to animals—often by injecting it into their brains—and made them sick. As a result, he became the celebrity scientist of his time, feted by kings and prime ministers, and hailed as a great scientist. His work led to pasteurization, a technique responsible for destroying the integrity and health-giving properties of milk (see chapter 9). His experiments ushered in the germ theory of disease, and for over a century this radical new theory has dominated not only the practice of Western medicine, but also our cultural and economic life.

We are proposing a different way of understanding the milk study. For example, what if the milk came from cows that were being poisoned or starved? Maybe they were dipped in flea poison; maybe they were fed grains sprayed with arsenic instead of their natural diet of grass; maybe they were fed distillery waste and cardboard—a common practice in Pasteur's day in many cities around the world.

We now know with certainty that any toxins fed to a nursing mammal show up in her milk. What if these listeria bacteria are not the cause of anything but simply nature's way of digesting and disposing of toxins? After all, this seems to be the role that bacteria play in biological life. If you put stinky stuff in your compost pile, the bacteria feed on the stuff and proliferate. No rational person would claim the compost pile has an infection. In fact, what the bacteria do in the compost pile is more of a bioremediation. Or, consider a pond that has become a dumping ground for poisons. The algae "see" the poison and digest it, returning the pond to a healthier state (as long as you stop poisoning the pond). Again, this is bioremediation, not infection.

If you take aerobic bacteria—bacteria that need oxygen—and put them in an anaerobic environment in which their oxygen supply is reduced, they often produce poisons. Clostridia is a family of bacteria that under healthy circumstances ferments carbohydrates in the lower bowel to produce important compounds like butyric acid; but under anaerobic conditions, this bacteria produces poisons that can cause botulism. It's the poisons, not the bacteria itself, that make people sick; or more fundamentally, it's the environment or terrain that cause the bacteria to create the poisons.

Isn't it possible that toxins in the milk—possibly because the cow is not well nourished and cannot easily get rid of the toxins—account for the presence of listeria (which is always present in our bodies, along with billions of other bacteria and particles called viruses)? The listeria is simply biodegrading the toxins that proliferate due to the unhealthy condition of the milk.

The central question then is how can we prove that the listeria, and not something toxic in the milk, is causing the diarrhea? The answer is the same as in the ping-pong ball example: feeding a healthy person the milk is like throwing the bucket with stones, ice, and (yes) a ping-pong ball at the wall; it proves nothing. You *must* isolate the ball—in this case, the listeria—and feed only this to the healthy person or animal to see what happens. This is what Pasteur claims to have done in his papers.

Pasteur passed his laboratory notebooks along to his heirs with the provision that they *never* made the notebooks public. However, his grandson, Louis Pasteur Vallery-Radot, who apparently didn't care for Pasteur much, donated the notebooks to the French national library, which published them. In 1995, Professor Gerald Geison of Princeton University published an analysis of these notebooks, which revealed that Pasteur had committed massive fraud in all his studies. For instance, when he said that he injected virulent anthrax spores into vaccinated and unvaccinated animals, he could trumpet the fact that the unvaccinated animals died, but that was because he also injected the unvaccinated animals with poisons.

In the notebooks, Pasteur states unequivocally that he was unable to transfer disease with a pure culture of bacteria (he obviously wasn't able to purify viruses at that time). In fact, the *only* way he could transfer disease was to either insert the whole infected tissue into another animal (he would sometimes inject ground-up brains of an animal into the brain of another animal to "prove" contagion) or resort to adding poisons to his culture, which he knew would cause the symptoms in the recipients.[5]

He admitted that the whole effort to prove contagion was a failure, resulting in his reputed deathbed confession: "The germ is nothing; the terrain is everything." In this case, *terrain* refers to the condition of the animal or person and whether the animal or person had suffered from poisoning or starvation.

Since Pasteur's day, no one has demonstrated experimentally the transmissibility of disease with pure cultures of bacteria or viruses. No one has bothered since Pasteur's time to throw a ping-pong ball at a wall and see what happens. Incredible as that may seem, we are sitting on a house of cards that has resulted in incalculable harm to humanity, the biosphere, and the geosphere of the Earth.

In chapters 2 and 3, we will examine cases in which bacteria or viruses were falsely accused of causing disease. Read on, dear friends; the ride has only started.

CHAPTER 2

ELECTRICITY AND DISEASE

The earliest "electricians" were not technicians who installed wires in houses; they were physicians and "healers" who used the newly discovered phenomena of electric current and static electricity to treat people with ailments—from deafness to headaches to paralysis. The only problem with having patients touch Leyden jars (a device that stores a high-voltage electric charge) or subject themselves to electric currents was that it sometimes caused harm and occasionally killed them.

One thing these early electrical experimenters noted was that people showed a range of sensitivity to electricity. According to Alexander von Humboldt, a Prussian scientist who (among other experiments) subjected himself and others to the shocks of electric eels, "It is observed that susceptibility to electrical irritation and electrical conductivity, differs as much from one individual to another, as the phenomena of living matter differ from those of dead material."[1]

These early studies captured the imagination of researchers; they began to realize that electric currents ran through the bodies of frogs and humans and that even plants were sensitive to electrical phenomena. After a 1749 earthquake in London, British physician William Stukeley concluded that electricity must play a role in earthquakes because the residents of London felt "pains in their joints, rheumatism, sickness, headache, pain in their back, hysteric and nervous disorders . . . exactly upon electrification, and to some it has proved fatal."[2]

As early as 1799, researchers puzzled over the cause of influenza, which appeared suddenly, often in diverse places at the same time, and could not be explained by contagion. In 1836, Heinrich Schweich, author of a book on influenza, noted that all physiological processes produce electricity and theorized that an electrical disturbance of the atmosphere may prevent the body from discharging it. He repeated the then-common belief that the accumulation of electricity in the body causes the symptoms of influenza.[3]

With the discovery of the sun's electrical nature, scientists have made some interesting observations. The period 1645–1715 is one that astronomers call the Maunder Minimum, when the sun was quiet; astronomers observed no sunspots during the time span, and the northern lights (aurora borealis) were nonexistent; in 1715, sunspots reappeared, as did the northern lights. Sunspot activity then increased, reaching a high in 1727. In 1728, influenza appeared in waves on every continent. Sunspot activities became more violent until they peaked in 1738, when physicians reported flu in both man and animals (including dogs, horses, and birds, especially sparrows). By some estimates, two million people perished during the ten-year pandemic.

These and other facts about the relationship of influenza to disturbances in electricity come from a remarkable book, *The Invisible Rainbow* by Arthur Firstenberg.[4] Firstenberg chronicles the history of electricity in the United States and throughout the world, and the outbreaks of illness that accompanied each step toward greater electrification. The first stage involved the installation of telegraph lines; by 1875, these formed a spiderweb over the earth totaling seven hundred thousand miles, with enough copper wire to encircle the globe almost thirty times. With it came a new disease called neurasthenia. Like those suffering today from "chronic fatigue syndrome," patients felt weak and exhausted and were unable to concentrate. They had headaches, dizziness, tinnitus, floaters in the eyes, racing pulse, pains in the heart region, and palpitations; they were depressed and had panic attacks. Dr. George Miller Beard and the medical community observed that the disease spread along the routes of railroads and telegraph lines; it often resembled the common cold or influenza and commonly seized people in the prime of life.[5]

In 1889, we mark the beginning of the modern electrical era and also of a deadly flu pandemic, which followed the advent of electricity throughout the globe. Said Firstenberg: "Influenza struck explosively and unpredictably, over and over in waves until early 1894. It was as if something fundamental had changed in the atmosphere."[6]

Physicians puzzled over influenza's capricious spread. For example, William Beveridge, author of a 1975 textbook on influenza, noted, "The English warship *Arachne* was cruising off the coast of Cuba 'without any contact with land.' No less than 114 men out of a crew of 149 fell ill with influenza and only later was it learnt that there had been outbreaks in Cuba at the same time."[7]

During World War I, governments on both sides of the conflict installed antennas, which eventually blanketed the earth with strong radio signals—and during the latter part of 1918, disaster struck. The Spanish flu afflicted a third of the world's population and killed about fifty million people, more than the Black Death of the fourteenth century. To stop the contagion, communities shut down schools, businesses, and theaters; people were ordered to wear masks and refrain from shaking hands.[8]

Those living on military bases, which bristled with antennas, were the most vulnerable. A common symptom was bleeding—from the nostrils, gums, ears, skin, stomach, intestines, uterus, kidneys, and brain. Many died of hemorrhage in the lungs, drowning in their own blood. Tests revealed a decreased ability of the blood to coagulate. Those close to death often developed "that peculiar blue color which seemed to mark all early fatal cases."[9]

Health officials were desperate to find a cause. The team of physicians from the US Public Health Service tried to infect their one hundred healthy volunteers at a naval facility on Gallops Island in Boston Harbor. A sense of frustration pervades the report, written by Milton J. Rosenau, MD, and published in the *Journal of the American Medical Association*.[10] Rosenau had built a successful career in public health by instilling a fear of germs, overseeing quarantines, and warning the public about the dangers of raw milk. He believed that something called *Pfeiffer bacillus* was the cause. The researchers carefully extracted throat and nasal mucus and even lung material from cadavers and transferred it to the throats, respiratory tracts, and noses of volunteers. "We used some billions of these organisms, according to our estimated counts, on each one of the volunteers, but none of them took sick," he said.

Then they drew blood from those who were sick and injected it into ten volunteers. "None of these took sick in any way."

Thoroughly perplexed, Rosenau and the other researchers designed the next experiment "to imitate the natural way in which influenza

spreads, at least the way in which we believe influenza spreads, and I have
no doubt it does [even though his experiments showed that it doesn't]—by
human contact." They instructed those afflicted to breathe and cough over
volunteers. "The volunteer was led up to the bedside of the patient; he
was introduced. He sat down alongside the bed of the patient. They shook
hands, and by instructions, he got as close as he conveniently could, and
they talked for five minutes. At the end of the five minutes, the patient
breathed out as hard as he could, while the volunteer, muzzle to muzzle (in
accordance with his instructions, about 2 inches between the two), received
this expired breath, and at the same time was breathing in as the patient
breathed out. This they repeated five times." The volunteers were watched
carefully for seven days, but alas, "none of them took sick in any way."

"Perhaps," said Rosenau, "there are factors, or a factor, in the transmission of influenza that we do not know. . . . Perhaps if we have learned
anything, it is that we are not quite sure what we know about the disease."

Researchers even tried to infect healthy horses with the mucous
secretions of horses with the flu[11]—yes, animals also became ill during
the pandemic—but the results were the same. The Spanish flu was not
contagious, and physicians could attach no blame to the accused bacterium nor provide an explanation for its global reach.

The year 1957 marked the installation of radar worldwide. The
"Asian" influenza pandemic began in February 1957 and lasted for a year.
A decade later, the United States launched twenty-eight satellites into the
Van Allen belts as part of the Initial Defense Communication Satellite
Program (IDCSP), ushering in the Hong Kong flu pandemic, which
began in July 1968.

As Firstenberg observed, "In each case—in 1889, 1918, 1957 and
1968—the electrical envelope of the earth . . . was suddenly and pro-
foundly disturbed,"[12] and along with it the electrical circuits in the
human body. Western medicine pays scant attention to the electrical
nature of living things—plants, animals, and humans—but mountains
of evidence indicate that faint currents govern everything that happens
in the body to keep us alive and healthy. From the coagulation of the
blood to energy production in the mitochondria, even to small amounts
of copper in the bones, which create currents for the maintenance of
bone structure—all can be influenced by the presence of electricity in the
atmosphere, especially "dirty" electricity, characterized by many overlapping frequencies and jagged changes in frequency and voltage. Today we
know that each cell in the body has its own electrical grid, maintained by

structured water inside the cell membrane (see chapter 8). Cancer occurs when this structure breaks down, and cancer has increased with each new development in the electrification of the earth.[13]

Humankind has lived for thousands of years with our brains tuned to the Schumann resonances of the earth, our bodies and indeed all life bathed in a static electric field of 130 volts per meter. The electronic symphony that gives us life is soft and delicate. Minute electrical currents that course through leaf veins or through the glial cells in our nervous system guide the growth and metabolism of all life-forms. Our cells communicate in whispers in the radiofrequency range.

Traditional Chinese medicine has long recognized the electrical nature of the human body and has developed a system to defuse the "accumulation of electricity" that leads to disease. It's called acupuncture. Many things that we do instinctively also help release any unhealthy buildup of current—the mother who strokes her infant's head or who scratches her children's backs to put them to sleep, the caresses of lovers, walking barefoot on the earth, massage, even handshakes and hugs—all now discouraged by the frowny faces of health authorities.

Fast-forward to the Internet and cell phone era. According to Firstenberg, the onset of cell phone service in 1996 resulted in greater levels of mortality in major cities like Los Angeles, New York, San Diego, and Boston.[14] Over the years, wireless signals at multiple frequencies have filled the atmosphere to a greater and greater extent, along with mysterious outbreaks like SARS and MERS.

Today the quiet hum of life-giving current is infiltrated by a jangle of overlapping and jarring frequencies—from power lines to the fridge to the cell phone. It started with the telegraph and progressed to worldwide electricity, then radar, then satellites that disrupt the ionosphere, then ubiquitous Wi-Fi. The most recent addition to this disturbing racket is fifth generation wireless—5G.

5G is broadcast in a range of microwave frequencies: mostly 24–72 GHz, with the range of 700–2500 MHz also considered 5G. Frequencies in this range (below the frequency of light) are called *nonionizing*, in contrast with *ionizing* radiation, which has a higher frequency than light. Ionizing radiation, such as X-rays, causes electrons to split off atoms, obviously something to which exposure must be limited. (This is why a lead shield is put on patients when they get X-rays.)

Instead of producing charged ions when passing through matter, nonionizing electromagnetic radiation changes the rotational, vibrational, and electronic valence configurations of molecules and atoms. This produces thermal effects (think microwave ovens). The telecommunications industry flatly denies any nonthermal effects on living tissue, even though a large body of research suggests considerable harm to the delicate electromagnetic systems in the human body from constant exposure to nonionizing frequencies. In particular, high-frequency electromagnetic fields like 5G affect cell membrane permeability[15]—not a good thing when the architecture of a healthy cell ensures that it is *not* permeable except in controlled situations.

We are already familiar with millimeter wave technology; this is the frequency of airport scanners, which can see through your clothes. Children and pregnant women are not required to go through these scanners, a nod to potential dangers. Adults get zapped a second or two; 5G bathes us in the same kind of radiation twenty-four seven.

Of particular concern is the fact that some 5G transmitters broadcast at 60 GHz, a frequency that is absorbed by oxygen, causing the oxygen molecule (composed of two oxygen atoms) to split apart, making it useless for respiration.[16]

On September 26, 2019, 5G wireless was turned on in Wuhan, China

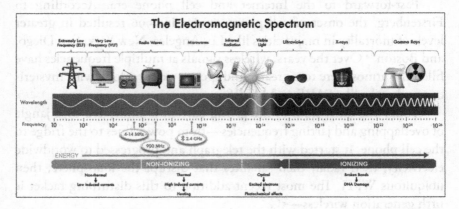

(and officially launched November 1) with a grid of about ten thousand 5G base stations—more than exist in the entire United States—all concentrated in one city.[17] A spike Covid in cases occurred on February 13—the same week that Wuhan turned on its 5G network for monitoring traffic.[18]

Illness has followed 5G installation in all the major cities in America, starting with New York in Fall 2019 in Manhattan, along with parts of Brooklyn, the Bronx, and Queens—all subsequent coronavirus hot spots. Los Angeles, Las Vegas, Dallas, Cleveland, and Atlanta soon followed, with some five thousand towns and cities now covered. Citizens of the small country of San Marino (the first country in the world to install 5G, in September 2018) have had the longest exposure to 5G and the highest infection rate—four times higher than Italy (which deployed 5G in June 2019), and twenty-seven times higher than Croatia, which has not deployed 5G.[19] In rural areas, the illness blamed on the coronavirus is slight to nonexistent.[20]

In Europe, illness is highly correlated with 5G rollout. For example, Milan and other areas in northern Italy have the densest 5G coverage, and northern Italy has twenty-two times as many coronavirus cases as Rome.[21]

In Switzerland, telecommunications companies have built more than two thousand antennas, but the Swiss have halted at least some of the 5G rollout due to health concerns. Switzerland has had far fewer coronavirus cases than nearby France, Spain, and Germany, where 5G is going full steam ahead.

Iran announced an official 5G launch in late March 2020, but assuming prelaunch testing in February, the advent of 5G correlates with the first Covid-19 cases at the same time. Korea has installed over seventy thousand 5G bases and reported over eight thousand cases of illness by mid-March. Japan began testing 5G in tunnels in Hokkaido in early February 2020, and this city now has the most cases of coronavirus in Japan, even more than Tokyo.[22]

In South America, the 5G rollout has occurred in Brazil, Chile, Ecuador, and Mexico, all of which have many coronavirus cases. Countries without 5G, such as Guyana, Suriname, French Guiana, and Paraguay have not reported any cases. Paraguay is doing what all countries should do—building a national fiber optics network without resorting to 5G.[23]

Bartomeu Payeras i Cifre, a Spanish epidemiologist, has charted the rollout of 5G in European cities and countries with cases per thousand people and demonstrated "a clear and close relationship between the rate of coronavirus infections and 5G antenna location."[24]

What about Covid-19 in the Amazon basin? The Pan American Health Organization (PAHO) estimates that there are at least twenty thousand active coronavirus cases among the indigenous peoples.[25] They live a primitive lifestyle, but 5G is already there,[26] along with "twenty-five enormously powerful surveillance radars, ten Doppler weather radars, two hundred floating water-monitoring stations, nine hundred radio-equipped 'listening posts,' thirty-two radio stations, eight airborne state-of-the-art surveillance jets equipped with fog-penetrating radar, and ninety-nine 'attack/trainer' support aircraft,' [all of] which can track individual human beings and 'hear a twig snap' anywhere in the Amazon."[27] These were installed in 2002 as part of the System for Vigilance of the Amazon (SIVAM), which monitors activities in a two-million-square-mile area of remote wilderness. All life in the Amazon is bathed with a range of electromagnetic frequencies.

These 5G frequencies go only a short distance and cannot penetrate into buildings. However, a few tech startups are working to get the 5G signal into the areas where we work, play, and sleep. Pivotal Commware is testing an "Echo 5G In-Building Penetration Device."[28] Pivotal's offices are about one mile from the Life Care nursing home in Kirkland, Washington, where the illness first appeared in the United States, and where twenty-five residents died. Was the Life Care center a testing ground for Pivotal's new device? Health-care facilities also teem with electronic equipment, some of it located right by the heads of sick patients. People who suffer from electrical hypersensitivity cannot go near many hospitals and nursing homes.

The 5G system is also installed on modern cruise ships. For example, the *Diamond Princess* cruise ship advertises "the best Wi-Fi at sea."[29] On February 3, 2020, the ship was quarantined in Yokohama, Japan after many passengers complained of illness. In the end, 381 passengers and crew members became sick, and fourteen died.

The *Diamond Princess* cruise ship. The four round objects on the top of the ship are 5G antennas and transmitters.

Of interest is the fact that the military has crowd-control devices that operate in the same ranges: 6–100 GHz. The 95 GHz Active Denial

System is a weapon that can penetrate the skin and produce intolerable heating sensations, causing people to move away from the beam.[30]

The EUROPA EMF Guideline 2016 states "there is strong evidence that long-term exposure to certain EMFs is a risk factor for diseases such as certain cancers, Alzheimer's disease, and male infertility. . . . Common EHS (electromagnetic hypersensitivity) symptoms include headaches, concentration difficulties, sleep problems, depression, lack of energy, fatigue, and *flu-like symptoms* [emphasis added]."[31]

An article published in May 2020 in *Toxicology Letters* found that in real-world conditions, exposure to wide-spectrum nonionizing frequencies adversely impacted skin, eyes, heart, liver, kidney, spleen, blood, and bone marrow.[32] Electromagnetic frequencies also disturb the immune function through stimulation of various allergic and inflammatory responses, and they adversely affect tissue repair.[33]

The Russians studied the effects of millimeter waves on animals and humans in 1979. Workers servicing ultra-high-frequency generators complained of fatigue, drowsiness, headaches, and loss of memory. The blood was particularly affected, with a reduction in the amount of hemoglobin and a tendency toward hyper-coagulation.[34] Even earlier, in 1971, the US Naval Medical Research Institute published more than twenty-three hundred references in a "Bibliography of Reported Biological Phenomena ('Effects') and Clinical Manifestations Attributed to Microwave and Radio-Frequency Radiation."[35] They found adverse effects almost everywhere in the body; in addition to "generalized degeneration of all body tissue," they noted altered sex ratio of births (more girls), altered fetal development, decreased lactation in nursing mothers, seizures, convulsions, anxiety, thyroid enlargement, decreased testosterone production, and—of particular interest—sparking between dental fillings and a peculiar metallic taste in the mouth.

One research review of almost two hundred studies[36] noted, "non-thermal effects have been clearly demonstrated in thousands of peer-reviewed publications." Whereas some EMF frequency band patterns are coherent and may be health-promoting, "the chosen 5G frequencies belong for a great part to the detrimental zones." The authors noted that government studies claiming 5G safety have taken no account of the fact that 5G radiation can be pulsating and modulated and emitted from multiple antennas. Of interest is the finding "that EMF waves can also be circularly polarized by interaction with atmospheric dust and therefore may penetrate much deeper into the organism. In addition, 5G waves may exhibit interference with other EMF wave frequencies, resulting in standing waves and environmental 'hot spots' of radiation that can be very taxing on EMF hypersensitive individuals." Air pollution and 5G are not a good mix!

A study published in *Frontiers in Oncology* describes lung injury from radiation therapy. Radiation therapy uses shorter waves at close range for a shorter period of time, but it stands to reason that 5G millimeter waves, with transmitters nearby, pulsing massive amounts of frequency at all times, could also cause lung injury. According to the authors, "Depending on the dose and volume of lung irradiated, acute radiation pneumonitis may develop, characterized by dry cough and dyspnea (shortness of breath)."[37]

Of interest is the fact that Lloyd's of London and other insurance carriers won't cover injury from cell phones, Wi-Fi, or smart meters. EMFs are classified as a pollutant, alongside smoke, chemicals, and asbestos: "The Electromagnetic Fields Exclusion (Exclusion 32) is a General Insurance Exclusion and is applied across the market as standard. The purpose of the exclusion is to exclude cover for illnesses caused by continuous, long-term non-ionizing radiation exposure i.e. through mobile phone usage."[38]

According to Dr. Cameron Kyle-Sidell, working in an emergency room (ER) in New York, the afflicted are literally gasping for air. "We've never seen anything like it!" he said.[39] Covid-19 patients' symptoms resemble those of high-altitude sickness rather than viral pneumonia. In fact, the ventilators that the hospitals have scrambled to obtain may do more harm than good and may be accounting for the high mortality rate, as they increase pressure on the lungs. These patients don't need help breathing—they need more oxygen when they take a breath. Many turn blue in the face. These are not signs of a contagious disease but of disruption of our mechanisms for producing energy and getting oxygen to the red blood cells.

Remember that during the Spanish flu, the problem was the lack of blood coagulability; with Covid-19, a key problem is lack of oxygen in the blood—both conditions point to electrical toxicity rather than infection—iron-rich blood cells would be especially vulnerable to the effects of electromagnetism.

And there's another symptom: fizzing. Many Covid patients report strange buzzing sensations throughout their body, "an electric feeling on the skin," or skin that feels like it is burning. Those who are electrically sensitive report similar sensations when they are near a cell phone or use GPS-guided cruise control in their cars. Other symptoms include a loss of smell and taste, fever, aches, breathlessness, fatigue, dry cough, diarrhea, strokes, and seizures—all of which are also reported by those who are electrically sensitive.

The correlation of 5G rollout and Covid-19 cases, and the similarity of symptoms, should give us pause. Shouldn't we look more closely before we institute mandatory vaccination and electronic ID chipping? Shouldn't we test to see whether this virus is actually contagious before we mandate social distancing and prescribe face masks?

Today's pandemic raises many questions. What makes some people more vulnerable than others to the effects of 5G? Why did thirty-five sailors on the battleship *Arachne* not get sick? Which environmental factors weaken our defenses? How should we treat this disease if it is not a viral disease? What about our diets? Can we protect ourselves with the right food choices? We will address these questions in subsequent chapters.

Most important, we will show that the minute particles called viruses are actually exosomes—not invaders but toxin-gobbling messengers that our cells produce to help us adjust to environmental assaults, including electro-smog. After all, most people have adjusted to worldwide radio waves, electricity in their homes, and ubiquitous Wi-Fi (and the sparrow population rebounded after the flu of 1738); exosomes are what allow this to happen. These tiny messengers provide real-time and rapid genetic adaptation to environmental changes. Whether these exosomes can help us adapt to the extreme disruption of 5G is the question of the day.

CHAPTER 3

PANDEMICS

Throughout history, philosophers believed that comets were "harbingers of doom, disease, and death, infecting men with a blood lust to war, contaminating crops, and dispersing disease and plague."[1]

The Chinese textbook *Mawangdui Silk* details twenty-nine types of comets, dating back to 1500 BC, and the disasters that followed each one. "Comets are vile stars," wrote a Chinese official in 648 AD. "Every time they appear in the south, they wipe out the old and establish the new. Fish grow sick, crops fail. Emperors and common people die, and men go to war. The people hate life and don't want to speak of it."[2]

In medieval Europe and even in colonial America, observers associated the appearance of comets with the onset of disease.[3]

In the summer of 536 AD, a mysterious and dramatic cloud of dust appeared over the Mediterranean and for eighteen months darkened the sky as far east as China. According to the Byzantine historian Procopius, "During this year a most dread portent took place. For the sun gave forth its light without brightness . . . and it seemed exceedingly like the sun in eclipse, for the beams it shed were not clear."[4]

Analysis of Greenland ice deposited between 533 and 540 AD shows high levels of tin, nickel, and iron oxides, suggesting that a comet or fragment of a comet may have hit the Earth at that time.[5] The impact likely triggered volcanic eruptions, which spewed more dust into the atmosphere. With the darkened sky, temperatures dropped, crops failed, and famine descended on many parts of the world.

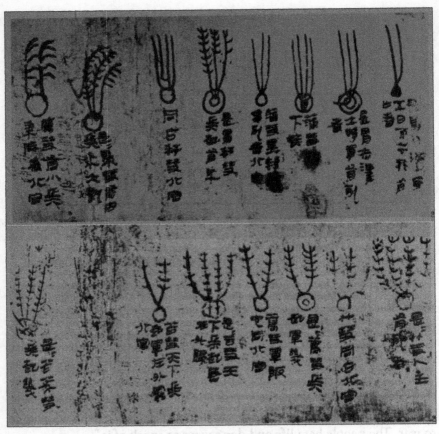

Depiction of various types of comets in Chinese documents.

Shortly afterward, in 541 AD, a mysterious illness began to appear on the outskirts of the Byzantine Empire. Victims suffered from delusions, nightmares, and fevers; they had lymph node swellings in the groin, armpits, and behind their ears. The plague, named after the reigning Emperor Justinian, arrived in Constantinople (the capital) in 542. Procopius noted that bodies were left stacked in the open due to a lack of space for proper burial. He estimated that in the city at its peak, the plague was killing ten thousand people per day.[6]

The current explanation for the correlation of comets and disease is that of "panspermia." We now know that outer space is populated by clouds of microorganisms, and the theory holds that comets are watery bodies—dirty snowballs—that rain new microscopic forms on the earth, to which humans and animals have no immunity.[7]

However, recent evidence indicates little if any water on comets. Rather, they are asteroids that have an elliptical orbit and become charged electrically as they approach the sun, an exchange that creates the comet's bright coma and tail. Their surfaces exhibit the kind of features that happen with intense electrical arcing, like craters and cliffs; bright or shiny spots on otherwise barren rocky surfaces indicate areas that are electrically charged. Comets contain mineral alloys requiring temperatures in the thousands of degrees, and they have sufficient energy to emit extreme UV light and even powerful X-rays. Moreover, as comets approach the sun, they can provoke high-energy discharges and flare-ups of solar plasma, which reach out to the comet.[8]

Bright spots on a comet's surface indicate strong electromagnetic activity.

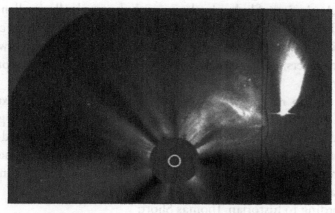

A solar flare reaches out to a highly charged comet.

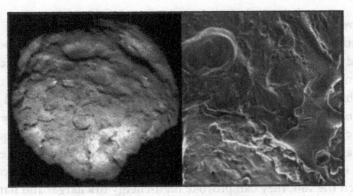

Cliffs and pitting on a comet's surface indicate intense electrical arcing.

Thus, comets can create electrical disturbances in the atmosphere even more powerful than those created by man-made electrification— and this radiation includes demonstrably dangerous *ionizing* radiation. No wonder the ancients were afraid of comets!

The conventional view holds that the Plague of Justinian was a case of bubonic plague. Researchers analyzed the remains from graves of the period and detected DNA from the *Yersinia pestis.*[9] Mainstream thinking has concluded that rats and other rodents carry *Yersinia pestis* and pass it along to fleas. When rats die, the blood-sucking fleas leave them to prey on other rats, dogs, and humans. The bacteria then enter humans via flea bites. Researchers believe that during the time of Justinian, rats on merchant ships carried the microorganism to the other Mediterranean ports.

The classic sign of bubonic plague are buboes—badly swollen lymph nodes. These often appear in the groin because, according to conventional thinking, most fleabites occur on the legs. Those infected will first experience fevers, chills, and muscle pains before developing septicemia or pneumonia.

The plague reappeared at periodic intervals over the next three hundred years, with the last recorded occurrence in 750 AD—possibly explained by still-orbiting cometary debris. It eventually claimed 25 percent of inhabitants in the Mediterranean region. Then the plague disappeared from Europe until the Black Death of the fourteenth century— also presaged by a comet.

According to historian Thomas Short:

In France . . . was seen the terrible Comet called Negra. In December
appeared over Avignon a Pillar of Fire. There were many great Earthquakes,
Tempests, Thunders and Lightnings, and thousands of People were swal-
lowed up; the Courses of Rivers were stopt; some Chasms of the Earth
sent forth Blood. Terrible Showers of Hail, each stone weighing 1 Pound
to 8; Abortions in all Countries; in Germany it rained Blood; in France
Blood gushed out of Graves of the Dead, and stained the Rivers crimson;
Comets, meteors, Fire-beams, coruscations in the Air, Mock-suns, the
Heavens on Fire.[10]

According to textbooks, the same bubonic plague organism of
Justinian's time caused the Black Death in Europe, 1347–1350. However,
some investigators have pointed out flaws in this theory. Although
researchers found evidence of *Yersina pestis* in dental pulp from a mass
grave of the period in France, other teams of scientists were unable to find
evidence of the pathogen in five other grave sites of the period from other
parts of Europe.[11]

Sociologist Susan Scott and biologist Christopher J. Duncan claim
that a hemorrhagic fever, similar to the Ebola virus, caused the Black
Death. Others blame anthrax or some now-extinct disease. They note that
medieval accounts don't square with modern descriptions of the illness.
Witnesses described a disease that spread at great speed with very high
mortality, unlike the plague, which moves slowly and had a death rate
of about 60 percent. Accounts describe buboes covering the entire body
rather than limited to the groin area as in the case of plague. Symptom
descriptions mention awful odors, splotches resembling bruises, delirium,
and stupor—none of which happen with modern-day bubonic plague.
Some critics have embraced the theory that a virus caused the disease,
but this premise hardly provides a better explanation than does bacteria
to explain the disease's rapid spread and high mortality.

Then there is the rat problem. No written documents from that time
describe vast legions of dead rats required to explain the plague. The
Black Death killed over half of Iceland's population, but rats didn't reach
Iceland until the nineteenth century. And the Black Death continued to
kill people during the winter months in northern Europe despite the fact
that the plague organism requires relatively warm temperatures.[12]

In *New Light on the Black Death: The Cosmic Connection*, Professor
Mike Baillie argues that a comet caused the pandemic. He points out

that witnesses of the period describe a significant earthquake on January 25, 1348, with other earthquakes to follow. "There have been masses of dead fish, animals, and other things along the sea shore and in many places covered in dust," wrote a contemporary observer. "And all these things seem to have come from the great corruption of the air and earth." Other documents describe tidal waves, rains of fire, foul odors, strange colors in the sky, mists and even dragons, in addition to earthquakes.[13]

Baillie believes that fragments from Comet Negra, which passed by earth in 1347, caused the atmospheric phenomena. Some fragments descended and injected huge amounts of dust into the atmosphere. Tree ring analysis indicates that as the material descended from space, it spewed large amounts of chemicals based on carbon and nitrogen into the stratosphere. According to Baillie, illness and death resulted from poisoned water and air as the comet flew overhead.[14]

But the symptoms—especially bruise-like blotches on the skin and high fatality rate—indicate radiation poisoning, probably rendered even more deadly by dust and ammonia-like compounds in the atmosphere. Imagine a large comet passing near the earth, crackling with intense electrical arcing, pelting the earth with X-rays and casting off fragments that fall to the earth and spew up toxic clouds of dust, followed immediately by horrible death, sometimes wiping out whole towns. This is not the kind of catastrophe that we can blame on microbes.

Perhaps our solar system is calming down—mankind has not seen such violent phenomena for centuries. But smaller electrical disturbances, ones that can't be seen, are still likely to promote outbreaks, albeit less disastrous. And if radiation poisoning—whether ionizing or nonionizing—provokes disease, there are obvious cofactors. Poisons in air, water, and food; toxins from insect bites; deadly fungi on grains; exposure to filth; malnutrition; and starvation; as well as fear and despair—we don't need to resort to the notion of contagion to explain outbreaks of disease.

Let us consider insect-borne diseases. Many (if not most) biting or stinging insects release toxins—often complex chemicals that can target the nervous system. Wasps, bees, flies, beetles, mosquitos,[15] ticks, bedbugs, lice, and ants all produce poisonous substances. Early studies suggest that insect saliva has chemicals with vasodilatory, anticoagulant, and immunosuppressive properties, although in recent times there has been little interest in (or research money for) the study of insect saliva.

In addition to overt poisons, insect saliva may contain parasite eggs. Tapeworms can be transmitted by fleas, and mosquito bites contain the eggs of plasmodium, a parasite said to cause malaria. Mosquitos also carry fly larvae, which can enter the body through bites, causing myiasis, a parasitic infestation of the body by fly larvae (maggots), which grow inside the host. Some mosquito species can carry filariasis, a parasite that causes a disfiguring condition called elephantiasis. These diseases are "infectious" in the sense that people acquire them from something outside the body, such as an insect, but only in the most bizarre of circumstances can they be transferred from one human being to another.

Actually, scientists have yet to solve the mystery of malaria, a disease that kills over one thousand people per day. The conventional view is that mosquitoes in tropical and subtropical regions transfer parasites to human blood through their bites, and this parasite then destroys red blood cells and causes intermittent fever. But the type of mosquito said to cause malaria inhabits every continent except Antarctica, including Europe and North America, where malaria is no longer a problem. From the fifteenth century until recent times, many people in England suffered from malaria under the name of "marsh fever" or "ague"—always associated with living in swampy marshes. In fact, what is common to areas known for malaria (both today and in the past) is human habitation in swamps and wetlands—and not just warm wetlands (which are conducive to mosquitos) but also wetlands in cooler areas such as England.

Wetlands produce swamp gases—a mixture of hydrogen sulfide, carbon dioxide, and especially methane. Methane poisoning causes fever, headaches, muscle weakness, nausea, vomiting, and feelings of asphyxiation—remarkably similar to the symptoms of malaria: fever, muscle weakness, nausea, vomiting, and chest and abdominal pain. Like malaria, methane poisoning can result in the destruction of red blood cells.[16] In areas of the world where people still live in swampy areas, intermittent exposure to swamp gases, which are undoubtedly stronger during warm weather or flooding seasons, seems a better explanation than mosquitos for this stubborn disease.

The conventional view holds that "viral diseases" such as yellow fever, dengue fever, Zika fever, and chikugunya are transmitted by mosquitos carrying viruses that "attach to and enter susceptible cells." According to textbooks, once these viruses enter the body and begin to replicate inside the cells, they are contagious and are spread from person to person through airborne droplets, sexual contact, eating food and drinking water contaminated with the virus, and even touching surfaces and bodily fluids contaminated with the virus. But we don't need the concepts of viruses and contagion to explain these diseases. Environments infested with fleas, mosquitos, lice, and other insects carrying toxins or parasites will result in many individuals, especially individuals with suboptimal nutrition, manifesting similar symptoms—an "outbreak" that requires no premise of person-to-person contact, only many people subject to the same stressors. For example, the "outbreak" of Zika "virus," blamed for a rash of babies born with tragically small heads, followed an experiment that gave prenatal ultrasounds of increasing intensity, duration, and frequency to thousands of pregnant women in Brazil.[17]

Toxins are powerful stressors. Sewage fumes contain a mixture of toxic gaseous compounds, such as hydrogen sulfide, carbon dioxide, methane, and ammonia. High concentrations of methane and carbon dioxide displace oxygen. In conditions of low oxygen, beneficial fermentative bacteria begin producing toxins instead of helpful compounds. Industrial chemicals in sewage can add to the adverse effects, especially if these toxins make their way into drinking water. In times past, these toxins included mercury, arsenic, and lead. Lead used for roofing, tanks, gutters, pipes, cables, and winemaking (and even added to recipes in Roman times) poisoned directly, through drinking water, or through the skin. Renaissance noblewomen wore makeup containing white lead ore, vinegar, arsenic, hydroxide, and carbonate, applied to the face over egg whites or a mercury foundation. Arsenic face powder was the crowning touch.[18] The price for the flawless complexion was paralysis, madness, and death.

Leather tanning contributed greatly to water pollution. Lime, tannin, animal dung, urine, alum, and arsenic were used in the process; the Industrial Revolution added toxic chromium solution to the mix. Production of red paint and dyes, metal extraction, and caustic soda production released mercury. Both mercury and arsenic were popular ingredients in medicines, and they no doubt carried off as many people as the diseases themselves.

The severe vomiting, diarrhea, dehydration, and muscle cramping of cholera is blamed on the bacterium *Vibrio cholerae*, either from sewage-tainted water or shellfish like oysters living in sewage-tainted water. Actually, the killer is a toxin—called "cholera toxin" (CT), which the bacteria produce under low-oxygen conditions. Although CT can be deadly, it also has anti-inflammatory properties and has shown promise as an immunotherapeutic drug.

Cholera affects up to five million people, mostly in third world countries, and causes over one hundred thousand deaths per year. Treatment includes oral rehydration therapy and zinc supplementation. Children are highly susceptible to CT, as are those who are malnourished or have lowered immunity. One strange observation is the fact that type O blood types are more likely to contract cholera.[19]

Even today, with the medical world's fixation on person-to-person transmission of disease and prevention through vaccination, health authorities agree that the solution to cholera is better sanitation. Cholera is rarely spread directly from person to person, but only through filthy drinking water.

An outbreak of cholera occurred in Soho, London, in 1854. According to Judith Summers in *Broad Street Pump Outbreak*, "by the middle of the [nineteenth] century, Soho had become an insanitary place of cowsheds, animal droppings, slaughterhouses, grease-boiling dens and primitive, decaying sewers. And underneath the floorboards of the overcrowded cellars lurked something even worse—a fetid sea of cesspits as old as the houses, and many of which had never been drained. It was only a matter of time before this hidden festering time-bomb exploded."[20]

The previous year, over ten thousand people died of cholera in England. The outbreak in Soho appeared suddenly: "Few families, rich or poor, were spared the loss of at least one member. Within a week, three-quarters of the residents had fled from their homes, leaving their shops shuttered, their houses locked and the streets deserted. Only those who could not afford to leave remained there. It was like the Great Plague all over again."

Dr. John Snow lived in the center of the outbreak and traced the source to a pump on the corner of Broad and Cambridge Streets, at the epicenter of the epidemic. "I found," he wrote afterward, "that nearly all the deaths had taken place within a short distance of the pump." In fact, in houses much nearer another pump, only ten deaths occurred—and of those, five victims had drunk the water from the Broad Street pump. Workers in a local brewery did not get sick—they drank beer provided as a perk of employment. Dr. Snow blamed the outbreak not on toxins but on "white, flocculent particles," which he observed under a microscope.[21]

Three decades later, Robert Koch tried injecting a culture of these white flocculent particles into animals, without succeeding in getting them sick—so cholera failed his second postulate. Cholera also failed his first postulate, as *Vibrio cholerae* appeared in both sick and healthy people.[22] Even so, he remained convinced that this bacillus was the cause of cholera—old ideas are difficult to dislodge even in the face of conflicting evidence.

It bears emphasis that all cities up to the nineteenth century were "fetid seas" of horse droppings, stinking manure piles, primitive water sanitation, toxic chemicals, crowded living conditions, loose pigs, and even raw sewage dumped from houses. Swill from inner-city breweries went to cows in inner-city confinement dairies, producing poisoned milk in conditions of unimaginable filth. The death rate among children born in these conditions was 50 percent. Officials blamed the death rate on the milk, which became the justification for pasteurization laws instituted one hundred years later.[23] By then, the problem had resolved itself with improved water and sewer systems, better living conditions, the advent of refrigeration, laws prohibiting inner-city breweries and dairies, and (most important) replacement of the horse with the car. Automobiles and buses brought in a different kind of pollution, but new technologies at least ensured that the water was finally clean. Much "infectious disease" cleared up, thanks not to doctors but rather to inventors and civil engineers.

One invention that made life safer was the washing machine, making it easier to keep clothes and bedding clean, especially as more and more dwellings had hot running water. Another invention was the vacuum cleaner, which helped keep living quarters free of bugs. (Window screens also helped.)

At the turn of the twentieth century, health officials considered smallpox to be highly infectious, but one physician disagreed. Dr. Charles A. R. Campbell of San Antonio, Texas, believed that smallpox was transmitted by the bites of bedbugs.

The modern official view holds that smallpox resulted from contact with a contagious virus—"Transmission occurred through inhalation of airborne *Variola virus*, usually droplets expressed from the oral, nasal, or pharyngeal mucosa of an infected person. It was transmitted from one person to another primarily through prolonged face-to-face contact with an infected person, usually within a distance of 1.8 m (6 feet), but could also be spread through direct contact with infected bodily fluids or contaminated objects (fomites) such as bedding or clothing . . . the infected person was contagious until the last smallpox scab fell off . . . Smallpox was not known to be transmitted by insects or animals."[24] Note that this description is written in the past tense—the official view is that smallpox has been conquered by vaccination, not by something as simple as getting rid of bedbugs.

Dr. Campbell ran a "pest house" for smallpox patients in San Antonio, where he tried hard to infect himself and others by "fomites" and direct face-to-face contact with infected persons:

> As even the air itself, without contact, is considered sufficient to convey this disease, and touching the clothes of a smallpox patient considered equivalent to contract it, I exposed myself with the same impunity as my pest-house keeper. . . . After numerous exposures, made in the ordinary manner, by going from house to house where the disease was . . . I have never conveyed this disease to my family, or to any of my patients or friends, although I did not disinfect myself or my clothes, nor take any precautions whatever, except to be sure that no bedbugs got about my clothing.
>
> Another one of my experiments was thoroughly to beat a rug in a room, only eight or ten feet square, from which had just been removed a smallpox patient. . . . I beat this rug in the room until the air was stifling, and remained therin for thirty minutes. This represented the respiratory as well as the digestive systems as accepted avenues of infection. . . . After inhaling the dust from that rug, I examined my sputum microscopically the following morning and found cotton and woolen fibres, pollen and comminuted manure, and also bacteria of many kinds.[25]

Although Dr. Campbell subsequently mingled with family, patients, and friends, none contracted smallpox. He repeated these experiments with others, failing to infect, even when in contact with patients covered in sores, but he always found bedbugs in the houses of those who contracted the disease.[26]

The British and American colonists used smallpox as a weapon against the Native Americans—they did it by giving them *blankets*, thus spreading the bedbug to the New World.

Campbell treated smallpox by administering sources of vitamin C:

> The most important observation on the medical aspect of this disease is the cachexia [bad condition] with which it is associated, and which is actually the soil requisite for its different degrees of virulence. I refer to the *scorbutic cachexia*. Among the lower classes of people this particular acquitted constitutional perversion of nutrition is most prevalent, primarily on account of their poverty, but also because of the fact that they care

little or nothing for fruits or vegetables . . . that it is more prevalent in winter when the anti-scorbutics are scarce and high priced; and finally, that the removal of this perversion of nutrition will so mitigate the virulence of this malady as positively to prevent the pitting or pocking of smallpox.

A failure of the fruit crop in any particularly large area is always followed the succeeding winter by the presence of smallpox.[27]

Dr. Campbell also applied himself to the elimination of mosquitoes by constructing huge bat houses—he was a great admirer of this strange winged creature and knew how to harness its help in the elimination of annoying insects, assumed to cause malaria.[28] Campbell was an inventive and colorful character, full of good ideas, yet hardly mentioned in medical journals or in histories of disease. Where's the glamor of a solution that involves clean beds and fresh fruit compared with the heroics of vaccination—smallpox vaccinations so toxic that health officials no longer recommend them.

Dr. Campbell's Municipal Bat-Roost, which eliminated mosquitos from San Antonio without the use of toxic chemicals.

Unlike the forgotten Dr. Campbell, Dr. Robert Koch is immortalized as the father of microbiology and the germ theory. Unable to prove that a microorganism caused cholera,[29] and in the case of rabies, knowing that Pasteur was unable to even find an organism,[30] Dr. Koch turned his attention to tuberculosis (TB). According to a historical article published in *World of Microbiology and Immunology*:

> In six months, Koch succeeded in isolating a bacillus from tissues of humans and animals infected with tuberculosis. In 1882, he published a paper declaring that this bacillus met his four conditions—that is, it was isolated from diseased animals, it was grown in a pure culture, it was transferred to a healthy animal who then developed the disease, and it was isolated from the animal infected by the cultured organism. When he presented his findings before the Physiological Society in Berlin on March 24, he held the audience spellbound, so logical and thorough was his delivery of this important finding. This day has come to be known as the day modern bacteriology was born.[31]

In 1905, Dr. Koch received the Nobel Prize for proving that TB was an infectious disease.

Except he didn't.

In fact, he could find an organism in infected tissue only by using special staining methods after the tissue was heated and dehydrated with alcohol. The stain was a toxic dye, methylene blue, and the solution he used contained another toxin—potassium hydroxide (lye). When he injected the organism stained with these poisons into animals, they got sick. But what caused the illness, the bacillus or the poisons?[32] And TB does not even satisfy Koch's first postulate. Only one person in ten who tests positive for TB actually develops the disease; those who don't are said to have "latent TB."

Even into the 1930s and 1940s, some scientists remained skeptical of the germ theory for TB—many still believed that the cause was genetic. An investigator who disputed both theories was the dentist Weston A. Price, author of the groundbreaking book *Nutrition and Physical Degeneration*.[33] During the 1930s and 1940s, he traveled around the globe to study the health of so-called "primitive peoples," living on ancestral diets. As a dentist, he naturally observed dental and facial formation and the presence or absence of tooth decay. He found fourteen groups in regions as diverse as the Swiss Alps, the Outer Hebrides, Alaska, South

America, Australia, and the South Seas in which every member of the tribe or village exhibited wide facial structure, naturally straight teeth, and freedom from tooth decay.

He also noted the absence of disease in these well-nourished groups. As soon as the "displacing foods of modern commerce" made inroads into a population, they became vulnerable to both chronic and "infectious disease," especially TB. The children born to those who adopted the Western diet of "sanitary" processed food—sugar, white flour, canned foods, and vegetable oils—were born with more narrow faces, crowded and crooked teeth, pinched nasal passages, narrow configuration of the birth canal, and less robust body formation.

Price rejected the notion that TB was inherited or caused by a microorganism, transmittable by droplets released into the air in the coughs and sneezes of the infected; he surmised that the root cause was a malformation of the lungs, similar to the narrowing of the facial structure and "dental deformities" in those born to parents eating processed foods. In a visit to a pediatric TB ward in Hawaii, he noted that every patient had dental deformities.[34] These dental deformities did not cause TB, of course, but Dr. Price believed that the same conditions that prevented the optimal formation of the facial bones also prevented optimal formation of the lungs. It was the dead and dying tissue in the lungs that attracted bacteria, nature's cleanup crew, and not the microorganism that caused the disease.

He noted that Swiss villagers living off their native diets of raw dairy products, sourdough rye bread, and some meat and organ meats had no TB—and this was a time when TB was the number-one killer in Switzerland and elsewhere.[35] Likewise, inhabitants of Lewis Island in the Outer Hebrides were free of TB. Their nutrient-dense diet consisted of seafood, including fish livers and fish liver oil, along with oat porridge and oatcakes. They lived in thatched houses that had no chimneys, living in close quarters with smoky, polluted air night and day; still they had no TB. When modern foods made their appearance, the situation changed, and TB took hold. Health workers blamed the smoky air of their cottages (not a microorganism!) and made them install chimneys, but to no avail. Only Weston A. Price was curious about the fact that the well-nourished islanders were immune, even when living in smoke-filled houses.[36]

Similarly, he observed that African tribesmen living on traditional foods seemed immune to the diseases in Africa, even though they went barefoot, drank unsanitary water, and lived in areas that swarmed with mosquitos.[37] Europeans visiting Africa needed to cover themselves completely and sleep under protective netting to avoid disease. Once the continent of Africa became "coca-colonized," these diseases proliferated among the Africans.

During the time of Dr. Price's research, it was not the so-called infectious diseases of Africa that struck terror in American minds, it was polio. According to health officials, the cause was an infectious virus. This virus didn't just make people (especially young people) sick; it occasionally left them crippled. Pictures of grown men in iron lungs and children wearing leg braces seared the national consciousness.

In the mid-1950s, physician Morton S. Biskind testified before Congress. Dr. Biskind's message was not what the legislators wanted to hear: polio was the result of central nervous system (CNS) poison, not a virus, and the chief CNS poison of the day was a chemical called dichlorodiphenyltrichloroethane, commonly known as DDT.[38] Used in World War II to control mosquitos said to cause malaria and typhus among civilians and troops, its inventor, Paul Herman Müller,[39] was awarded the Nobel Prize in Physiology or Medicine in 1948 "for his discovery of the high efficiency of DDT as a contact poison against several anthropods."

By October 1945, DDT was available for public sale in the United States. Government and industry promoted its use as an agricultural and household pesticide—really promoted it. Photographs from the era show housewives filling their houses with DDT fog; dairy farmers dusting cows in their cowsheds, even spraying it into the milk; crop dusters depositing DDT on fields and forests; and children on beaches enveloped in the pesticide. An attachment for your mower could distribute DDT over your lawn, and trucks sprayed DDT on city streets, children cheerfully playing in the spray.

DDT largely replaced another CNS poison—lead arsenate, introduced in 1898 for use on crops and orchards. Before that, the preferred spray was plain arsenic. Biskind wrote:

> In 1945, against the advice of investigators who had studied the pharmacology of the compound and found it dangerous for all forms of life, DDT . . . was released in the United States and other countries for general use by the public as an insecticide. . . . It was even known by 1945 that DDT is stored in the body fat of mammals and appears in the milk. With this foreknowledge the series of catastrophic events that followed the most intensive campaign of mass poisoning in known human history, should not have surprised the experts. Yet, far from admitting a causal relationship so obvious that in any other field of biology it would be instantly accepted, virtually the entire apparatus of communication, lay and scientific alike, has been devoted to denying, concealing, suppressing, distorting and attempts to convert into its opposite, the overwhelming evidence. Libel, slander and economic boycott have not been overlooked in this campaign. . . .
>
> Early in 1949, as a result of studies during the previous year, the author published reports implicating DDT preparations in the syndrome widely attributed to a 'virus-X' in man, in 'X-disease' in cattle and in often fatal syndromes in dogs and cats. The relationship was promptly denied by government officials, who provided no evidence to contest the author's observations but relied solely on the prestige of government authority and sheer numbers of experts to bolster their position. . . .
>
> ['X-disease'] . . . studied by the author following known exposure to DDT and related compounds and over and over again in the same patients, each time following known exposure. We have described the syndrome as follows: . . . . In acute exacerbations, mild clonic convulsions involving mainly the legs, have been observed. Several young children exposed to DDT developed a limp lasting from 2 or 3 days to a week or more. . . .
>
> Particularly relevant to recent aspects of this problem are neglected studies by Lillie and his collaborators of the National Institutes of Health, published in 1944 and 1947 respectively, which showed that DDT may produce degeneration of the anterior horn cells of the spinal cord in animals. These changes do not occur regularly in exposed animals any more than they do in human beings, but they do appear often enough to be significant.

> When the population is exposed to a chemical agent known to produce in animals lesions in the spinal cord resembling those in human polio, and thereafter the latter disease increases sharply in incidence and maintains its epidemic character year after year, is it unreasonable to suspect an etiologic relationship?[40]

Investigator Jim West unearthed Biskind's writings and testimony, along with other reports about the effects of poisons on the CNS, dating from the mid-nineteenth century. West compiled the following graphs, noting the correlation of pesticide use and polio incidence in the United States.[41]

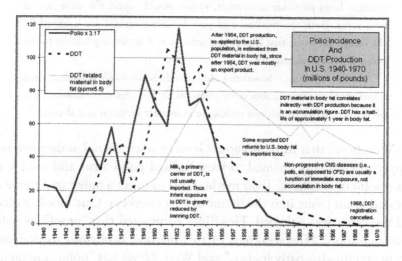

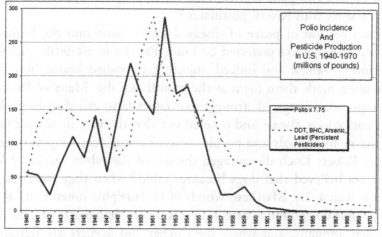

As use of DDT in the United States declined, so did the incidence of polio. Vaccination programs were introduced at the same time and take credit for the decline.

West says:

> A clear, direct, one-to-one relationship between pesticides and polio over a period of thirty years, with pesticides preceding polio incidence in the context of the [central nervous system]-related physiology . . . leaves little room for complicated virus arguments, even as a cofactor, unless there exists a rigorous proof for virus causation. Polio shows no movement independent from pesticide movement, as one would expect if it were caused by a virus. Both the medical and popular imaginations are haunted by the image of a virus that invades (or infects) and begins replicating to the point of producing disease.
>
> In the laboratory, however, poliovirus does not easily behave in such a predatory manner. Laboratory attempts to demonstrate causation are performed under conditions which are extremely artificial and aberrant.[42]

West notes that in 1908–1909, German researchers Landsteiner and Popper in Germany claimed to have isolated polio virus and used it to cause polio in monkeys. Their method was to inject a pulverized purée of diseased brain tissue into the brains of two monkeys. One monkey died, and the other was sickened. Headlines trumpeted this "proof" of polio virus causation. "The weakness of this method is obvious to everyone except certain viro-pathologists," said West. Never has "polio contagion" passed muster with Rivers' postulates.[43]

The injection of purée of diseased brain tissue into the brains of dogs was the method preferred by Louis Pasteur to establish microbial causation of rabies; and indeed, injecting smooshed brains into their heads often made them foam at the mouth and die. Many of Pasteur's contemporaries disagreed strongly that rabies (also called hydrophobia) was a contagious disease and pointed out that the vaccine often caused great harm to animals and people—even Pasteur's contemporary germ theorist Robert Koch discouraged the use of the rabies vaccine.[44] Vets of the era believed that dogs became "rabid" when they were starved and mistreated. Dr. Matthew Woods of Philadelphia noted that "at the Philadelphia dog pound, where on average more than 6,000 vagrant dogs are taken annually, and where the catcher and keepers are frequently bitten while handling them, not one case of hydrophobia [rabies] has occurred during its entire history of twenty-five years, in which time 150,000 dogs have been handled."[45] During the 1960s, researchers succeeded in inducing symptoms of rabies in experimental animals by putting them in bat caverns where they could breathe the toxic stultifying vapors of bat guano, later claiming to have isolated an "airborne rabies virus." To test whether this so-called "virus" caused rabies, one researcher "inoculated [mice] intracerebrally." Fifty percent died within forty-eight hours, but none developed rabies.[46]

As for polio, even with worldwide vaccination programs, polio has not gone away, either in the United States or in third world countries. Today in the United States, it has received a new name—acute flaccid paralysis (AFP), displaying symptoms identical to polio—with over two hundred cases recorded in 2018. Many parents have observed that the condition appears after a vaccination. CDC's pathetic advice: "To prevent infections in general, persons should stay home if they are ill, wash their hands often with soap and water, avoid close contact (such as touching and shaking hands) with those who are ill, and clean and disinfect frequently touched surfaces."[47]

In some areas of the world, such as India and Africa, the incidence of acute flaccid paralysis has skyrocketed, which many blame on campaigns to administer experimental polio vaccines to children ages zero to five.

Indian researchers described this strong correlation in a 2018 publication in the *International Journal of Environmental Research and Public Health* and calculated that, countrywide from 2000 to 2017, there were "an additional 491,000 paralyzed children" in excess of "the expected numbers."[48] Dr. Suzanne Humphries suggests that—far from credit for eliminating childhood paralysis going to vaccination campaigns—"there is strong evidence pointing to the likelihood that experimental polio vaccination is related to the sharp rise in AFP."[49]

If the true cause of epidemics is exposure to electrical pollution or toxins (from insects, industrial poisons, toxins produced by bacteria under conditions of filth, vaccinations, and drugs), with substandard nutrition as a cofactor, what about the outbreaks of disease in the Americas, in Africa, and in the South Seas, when these aboriginal peoples first met the European colonists? Didn't they begin to suffer from infectious disease as soon as they came in contact with infectious diseases carried to the New World on boats from the Old World—diseases to which they had no immunity?

Actually, native peoples did *not* contract disease immediately on contact with the Europeans. For example, fishermen and early explorers visited the northeastern waters along the Atlantic coast during the fifteenth and sixteenth centuries, yet we have no historical commentary on the existence of disease or epidemics among the aboriginal peoples during that time. According to Raymond Obomsawin, in his report "Historical and Scientific Perspectives on the Health of Canada's first Peoples,[50] "Since the prime purpose of this early contact was to commercially exploit natural resources, any visible evidence of the physical weakness or sickness of the indigenous inhabitants would surely have excited some keen interest." Instead, these early reports marveled at the Native Americans' good health and robust constitution.

Obomsawin notes that the first recorded outbreaks of disease in Native Americans living in the Ottawa Valleys occurred between 1734 and 1741. Samuel de Champlain had established the first European settlement at Quebec on the St. Lawrence River over one hundred years earlier, in 1608, and it wasn't until the 1800s that smallpox, dysentery, typhus, yellow fever, tuberculosis, syphilis, and various other "fevers" became prevalent in the aboriginal population.

By the mid-eighteenth century, Native American life had succumbed to serious disruptions. As a result of intensive trapping, the game populations had dwindled, seriously affecting the availability of food and skins to make clothing and footwear. During this period, sugar, white flour, coffee, tea, and alcohol arrived on trading ships, which the colonists traded with the Indians for furs.

The same pattern prevailed on the West Coast, where the salmon fisheries became depleted by the mid-1800s. These northwest peoples spoke of "disease boats" or "pestilence canoes," the Spanish and British seagoing vessels that arrived with increasing frequency. They brought smallpox, but also the foods that made them vulnerable to smallpox. An early one-hundred-foot sailing cargo vessel could transport as much as eight hundred thousand pounds of "goods"—or maybe we should say "bads"—including blankets for the Native Americans.[51]

Tribal peoples largely dependent upon the buffalo were not affected until the early 1870s, when the animals became depleted through exploitation and deliberate campaigns to kill off the herds upon which they depended.

According to a Canadian government report:

> The transformation of Aboriginal people from the state of good health
> that had impressed travelers from Europe to one of ill health . . . grew

worse as sources of food and clothing from the land declined and traditional economies collapsed. It grew worse still as once mobile peoples
were confined to small plots of land where resources and opportunities
for natural sanitation were limited. It worsened yet again as long-standing
norms, values, social systems and spiritual practices were undermined or
outlawed.[52]

Regarding the Plymouth colony, the Pilgrims were not the first
Europeans in the area. European fishermen had been sailing off the New
England coast, with considerable Native American contact, for much
of the sixteenth and seventeenth centuries, and trading for beaver skins
commenced in the early 1600s, prior to the arrival of the Pilgrims in 1620.

In 1605, the Frenchman Samuel de Champlain made an extensive and detailed map of the area and the surrounding lands, showing
the Patuxet village (site of the future town of Plymouth) as a thriving
settlement.

In 1617–1618, just prior to the arrival of the *Mayflower*, a mysterious
epidemic wiped out up to 90 percent of the Indian population along the
Massachusetts coast. History books blame the epidemic on smallpox, but
a recent analysis has concluded that it may have been a disease called leptospirosis.[53] Even today, leptospirosis kills almost sixty thousand people
per year.

Leptospirosis is a blood infection similar to malaria, associated with
various forms of spirochaete bacteria. Other forms of spirochete parasites characterize syphilis, yaws, and Lyme disease. Humans encounter
these spirochetes through animal urine or water and soil contaminated
with animal urine coming into contact with the eyes, mouth, nose,
or cuts. The disease is associated with poor sanitation. Both wild and

Native Americans trade
beaver skins with European
colonists for liquor and
other items that made them
vulnerable to disease.

domestic animals can transmit leptospirosis through their urine and other
fluids; rodents are the most common vector, and the beaver is a rodent.

One important factor omitted from discussions about Native
American diseases is the disruption of the salt trade. The first European
explorers in the New World did not come to the East Coast but to
Florida and the southeastern part of North America. During the 1540s,
eighty years before the Pilgrims landed at Plymouth Rock, the explorer
Hernando de Soto led the first European expedition deep into the terri-
tory of the modern-day United States. They traversed Florida, Georgia,
Alabama, and possibly Arkansas, and they saw the Mississippi River.

Some anthropologists have insisted that the Native Americans did
not consume salt, but de Soto received "an abundance of good salt"
as a gift from the Native Americans, and he observed the production
and trade of salt in the southeastern part of the country. In the lower
Mississippi Valley, he met traveling Native American merchants selling
salt. According to the de Soto records, lack of salt could lead to a most
unfortunate death:

> Some of those whose constitutions must have demanded salt more than
> others died a most unusual death for lack of it. They were seized with a
> very slow fever, on the third or fourth day of which there was no one at
> fifty feet could endure the stench of their bodies, it being more offensive
> than that of the carcasses of dogs or cats. Thus they perished without
> remedy, for they were ignorant as to what their malady might be or what
> could be done for them since they had neither physicians nor medicines.
> And it was believed that they could not have benefited from such had they
> possessed them because from the moment they first felt the fever, their
> bodies were already in a state of decomposition. Indeed, from the chest
> down, their bellies and intestines were as green as grass.[54]

The most important sources of salt were the salt springs that dotted
northwestern Louisiana, western Arkansas, and the Ohio River Valley.
Archeological remains in these areas indicate that the Native Americans
evaporated the brine in shallow clay salt pans, most likely by adding
hot rocks to the brackish water. They also retrieved salt from ashes of
certain plants and from salt-impregnated sand; they sometimes gathered
rock salt. Well-defined salt trails allowed the transport of salt to the east.
Coastal Native Americans generally got their salt through trade rather

than the evaporation of seawater, as wood for fire making is sparse near ocean beaches, and the moist sea air is unconducive to evaporation.[55]

The salt traders did not belong to any tribal group but traveled alone from tribe to tribe carrying baskets of salt gathered from salt lakes, along with other goods. As Native American cultural life crumbled in the face of the European invasion, the salt trade would have been an early victim of this disruption. Salt is critical for protection against parasites. We need the chloride in salt to make hydrochloric acid; without salt, the stomach will not be sufficiently acidic to kill parasites.

The point is that the so-called "infectious" diseases that caused so much suffering did not arrive until after a period of disruption and nutritional decline; and fear and despair almost certainly played a role. When disease broke out in a village, the afflicted often found themselves abandoned by those still healthy, so they had no one to care for them. Unable to get water for themselves, they typically died of thirst.[56] This may explain why the death rates during outbreaks were so much higher for the Native Americans (typically 90 percent) than for Europeans (typically 30 percent).

One disease blamed for Native American death was measles, considered to be a viral disease. But on February 16, 2016, the Federal Supreme Court of Germany (BGH) made a historic ruling: there is no evidence for the existence of a measles virus. The case grew out of a challenge by German biologist Stefan Lanka, who offered a sum of one hundred thousand Euros to anyone who could supply proof that the measles virus existed. A young doctor, David Bardens, took up the challenge, providing Lanka with six studies as proof that the measles virus did indeed exist. When Lanka claimed that these studies did not meet the evidence needed to claim the prize, Bardens took him to court. The court sided with Bardens and ordered payment of the prize.

But Lanka took the cases to the Supreme Court, where the judge decided in his favor and ordered the plaintiff to bear all the procedural costs. Lanka was able to show that the six studies misinterpreted "ordinary constituents of cells" as part of the suspected measles virus.[57]

According to Lanka, decades of consensus-building processes have created a model of a measles virus that doesn't actually exist: "To this day, an actual structure that corresponds to this model has been found neither in a human, nor in an animal. With the results of the genetic tests, all thesis of existence of measles virus has been scientifically disproved."[58]

The existence of a contagious measles virus justified the development of the measles vaccine, which has earned the pharmaceutical industry billions of dollars over a forty-year period. But if such a microorganism does not exist, said Lanka, "This raises the question of what was actually injected into millions of German citizens over the past decades. According to the judgment by the Supreme Court, it may not have been a vaccine against measles."[59]

But what about measles parties? What about successful attempts by parents to infect their children with the common childhood diseases like measles, chicken pox, and mumps? And what about sexually transmitted diseases (STD) like syphilis—said to be a disease that the Europeans contracted from the Native Americans? The mystery of childhood illnesses and sexually transmitted diseases will be addressed in chapter 7.

CHAPTER 4

FROM AIDS TO COVID

To understand the story of coronavirus, it's instructive to go back to 1971. That was the year that President Nixon declared the War on Cancer. The theory of the day was that viruses caused cancer, and the medical establishment vowed they would find a cure by 1975. Hundreds of millions of dollars "flowed into a completely one-sided cancer research focused on wonder-drug production."[1]

Of course, the war was a complete failure, and by the early 1980s, NIH scientists were scrambling to obtain continued funding and the CDC "needed a major epidemic to justify its existence."[2]

Enter AIDS (Acquired Immune Deficiency Syndrome), said to be caused by HIV (Human Immunodeficiency Virus), a deadly virus that "made the jump from other primates to humans in west-central Africa in the early-to-mid [twentieth] century."[3] Some predicted that as an STD, AIDS would eventually spread through the whole population and kill us all. Between 1981 and 2006, almost two hundred *billion* dollars was invested in AIDS research, focused on the virus hypothesis and development of toxic antiviral drugs like AZT.

Many books[4] have carefully documented the case against the claim that a virus called HIV causes the disease called AIDS. Unfortunately, these careful arguments seem to make no difference to the man on the street—nor to scientists, for that matter. In essence, no matter what the evidence, 99 percent of the population, including most of the alternative medical community, still believe in this myth.

Let's look at the facts. AIDS was not a disease new to the 1970s and early 1980s. In fact, it is the same disease as the condition caused by immune-suppressive drugs used to prevent people from rejecting organs such as a heart or kidney after a transplant. The only thing new about the disease was the appearance of a new type of cancer called Kaposi's sarcoma. AIDS is not a specific disease at all. It simply means a collapse of the cell-mediated immune system, known, even in the 1970s, to have many diverse causes. With the collapse of the immune system, you see conditions like frequent infections, TB, mononucleosis, peripheral neuropathy, and Guillain-Barré syndrome—all of which often come under the rubric of AIDS.

The new part of the disease, Kaposi's sarcoma, has been definitively linked to the use of "poppers" (alkyl nitrite drugs), which are immune-suppressive. This drug was used to relax the anal sphincter and facilitate anal intercourse. The vast majority of people who got Kaposi's sarcoma were gay men who used poppers (never people who "acquired the virus" from any other source). Once the use of poppers stopped in the gay community, so did Kaposi's sarcoma.

In spite of forty years of research, no one has isolated an HIV virus from any bodily fluid of a person suffering from AIDS. Not once. This is shocking for most people to hear, but cash awards are available for anyone who can show with an electron microscope purified HIV virus isolated from anyone with AIDS. Up to this point, no one has collected these cash awards.

No one has ever documented transmission of any purified HIV virus from one person or one animal to another with any resultant sickness. Not once. In fact, the biggest study on AIDS ever carried out[5] clearly showed that HIV is *not* transmissible through sexual contact.

And, finally, as we will discuss in chapter 5, the test used to make a "diagnosis" of AIDS can *never* determine causation. It is simply a test looking for genetic material of unknown origin. Since we have no proof that any virus or bacteria has ever caused any disease, the test is simply irrelevant for determining causation.

When the test for AIDS—called the PCR test—finds a higher level of genetic particles in the blood, it simply means that the person's condition is causing a lot of genetic deterioration—from toxins, EMF poisoning, malnutrition, or stress. The test can never determine the cause of the illness. If one first isolated, purified, and characterized the entire genome of the virus in question, then one could determine whether the snippet of genetic material you are looking for is unique to that virus in question. Absent doing a purification, isolation, and characterization step, there is simply no way to say that the snippet you are looking at is either unique to that virus or even originated in that virus.

If you poison an organism with any type of toxin that degrades your cells (which most poisons do, including EMF poisons[6]), then more genetic material will be found in your blood and the PCR test will pick this up. This means you are sick. This also applies to antibodies: the more poisoned you are, the more antibodies you tend to produce to protect yourself. This simple fact explains why all PCR and antibody tests, including those for HIV and the coronavirus, tend to show higher "viral loads" (which are not viruses but genetic material) and to be more positive in sicker people. It doesn't mean they have a viral infection; it means they are sick. This is why the package insert for PCR and antibody tests for both HIV and coronavirus say that you may get a false positive if the person has one of about forty conditions. These include strep throat, "viral infections," autoimmune disease, cancer, pregnancy, or nursing. In other words, any stress on the body provokes us to make more antibodies and have more degraded genetic material in our blood and other fluids— no surprise there. There is nothing in these tests that either proves viral causation or, absent purification, proves that the PCR snippet even came from the virus in question—*nothing*. It is simply a house of cards. (For more on testing, see chapter 5.)

Since these facts are obvious and easily proven, how can they have escaped the scrutiny of the "brilliant" men and women who run our health-care system and populate the ranks of virologists?

Once our health authorities declared that the multifactorial condition called "AIDS" was caused by a virus, they needed to come up with a way to treat it. The pharmaceutical companies (in particular, Burroughs-Wellcome), dusted off an old and very toxic drug called azidothymidine (AZT) and remarketed it for use with AIDS patients.

AZT is a nucleoside analogue drug; it interferes with the production of DNA from the RNA supposedly contained in the HIV virus. The theory is that without the ability of the HIV virus to make copies of DNA, it can't grow, replicate, and cause infection and disease. In practice, not only did AZT show much of the toxicity associated with cancer chemotherapy drugs (many of which also interfere with some aspect of DNA replication), it was shown to be worse than worthless in preventing the progression from asymptomatic to full-blown AIDS.

In the first trial of AZT on asymptomatic HIV-positive people, AZT was given to 877 people, while 872 received a placebo. As soon as a patient developed any AIDS symptoms, he or she (15 percent were women) would be offered "open label" AZT. The mortality rates were shocking; over the three years of the trial, there were seventy-nine AIDS-related deaths in the AZT group, but only 67 patients in the placebo group. So not only were you more likely to develop symptomatic AIDS if you took AZT compared to doing nothing, you were also subjected to all the usual toxicity of taking a chemotherapy drug.[7]

Adverse effects of AZT include nausea, vomiting, acid reflux (heartburn), headache, sleep problems, and loss of appetite. Long-term adverse effects include anemia, neutropenia (low white blood cells), hepatotoxicity (liver toxicity), cardiomyopathy (problems with the heart muscle), and myopathy (muscle weakness).

There are many parallels with today's situation in which "antiviral" drugs actually increase the likelihood of having a bad outcome from this imaginary viral disease called coronavirus.

In the wake of AIDS followed other "viral" diseases, including hepatitis C, SARS (Severe Acute Respiratory Syndrome), MERS (Middle East Respiratory Syndrome), bird flu, swine flu, Ebola, and Zika. Great sums were dedicated to finding viral causes and one-drug-fits-all cures, all following a familiar playbook:

> inventing the risk of a disastrous epidemic, incriminating an elusive pathogen, ignoring alternative toxic causes, manipulating epidemiology with non-verifiable numbers to maximize the false perception of an imminent catastrophe, and promising salvation with vaccines. This guarantees large financial returns. But how is it possible to achieve all of this? Simply by relying on the most powerful activator of human decision-making process—FEAR![8]

Needless to say, researchers have yet to prove that a virus causes any of these conditions.

Fast-forward to November 2019, when authorities in China noticed that a group of people were getting sick in a new way. They noticed that many of the people who contracted this illness reported visiting a particular fish market in Wuhan, China. The symptoms were respiratory in nature, including dry cough. Of course, these symptoms are not altogether new, as people throughout history have suffered from a variety of respiratory diseases such as bronchitis, pneumonia, and asthma. Still, as the number of cases increased, authorities looked into the situation. There was a suspicion that a new disease had appeared, which naturally triggered a search for the cause.

Because the symptoms of the sick people resembled pneumonia, some of the original patients received antibiotics. This was done because one of the recent "postulates" proving causation of an infectious disease states that if antibiotics fail to resolve the symptoms, this constitutes "presumptive" evidence (rather than "direct" evidence) that the pneumonia is caused by a virus (which obviously doesn't respond to antibiotic therapy). Since the patients didn't improve with antibiotic therapy, this triggered the hypothesis that the new type of pneumonia must be caused by a new or modified virus.

Let's consider our original ping-pong ball example given in chapter 1. The whole issue at this point revolves around whether the viral causation of this new set of symptoms can be proven. We want and should demand proof of causation, not a ping-pong ball in a bucket of stones and ice cubes, a computer simulation, an analysis of the virus, or the testimony of experts. We are betting our lives, our children, the world economy, and much more on the weight of the evidence. We need rock-solid proof; we need to follow the Koch's or Rivers' postulates, which entail a simple line of reasoning that any rational human being can recognize as the way to prove infectious causation. In other words, here is what *should* have happened in early 2020.

As soon as the Chinese medical authorities suspected an outbreak of a new and dangerous disease, they should have collected about five hundred people with identical, or at least nearly identical, symptoms. Then they should have identified another group of equal size as matched controls—that is, people of a similar age, lifestyle and background, also from Wuhan, who had no symptoms. Given the possible slow development of this illness, it would have been prudent to follow the five hundred control people over at least a few months to make sure that none developed these new symptoms.

The next step would be to do a thorough microbiological examination of a variety of fluids taken from these one thousand subjects. At a minimum, this should have included blood, sputum, urine, and nasal swabs. The examination should have included conventional light microscopy to look for bacteria and electron microscopy to look for viruses.

Once a novel bacteria or virus was found in *all* the sick people and *none* of the well people, the bacteria or virus should have been meticulously isolated, purified, and cultured in a neutral medium (which is actually not possible for viruses, since they "grow" only in other living cells). Once this purification step was accomplished, the purified microbe should have been introduced into a test animal in the normal way that one suspected the microbe might be spreading—by airborne droplets—*not* injected directly into the brain of the animal as scientists like Pasteur did to "prove" the contagious etiology of rabies, TB, or polio.

Finally, a control group of test animals should have been subjected to the same attempts at contamination. In other words, if you are going to spray purified virus into the animals' nostrils to see whether they get sick, you need to spray pure saline into the nostrils of a control group of animals to make sure the animals are not getting sick just because you are spraying stuff up their noses.

Any sane and logical person would agree that this is what should have happened. Finally, if for some reason the medical authorities in China were unable to carry out such an investigation, they should have enlisted the help of the CDC and the equivalent organizations in Europe and Russia, or the World Health Organization (WHO), to make sure the investigations were done carefully, properly, and thoroughly.

The amazing part of this story is that not only do we lack this kind of evidence for a viral cause of Covid-19, we also lack this kind of evidence for the many "viral" epidemics we have faced during the last century, including polio, AIDS, SARS, Ebola, Zika, and hepatitis C. In fact, not a single part of this clear and simple proof has been attempted.

Let's look then at what was done to prove that the severe acute respiratory syndrome coronavirus 2 (SARS-CoV-2) was the cause of this new set of symptoms. Four papers published in China are cited as proof that the new and novel coronavirus is the probable cause of this new disease.[9] For an in-depth analysis of these papers, please refer to a presentation by Andrew Kaufman MD,[10] in which he dissects in great detail the methods and conclusions of these seminal studies.

To review these four studies, let's look again at Rivers' postulates for determining whether a particular virus causes a disease.

1. The virus can be isolated from diseased hosts.
2. The virus can be cultivated in host cells.

3. Proof of filterability—the virus can be filtered from a medium that also contained bacteria.
4. The filtered virus will produce a comparable disease when the cultivated virus is used to infect experimental animals.
5. The virus can be re-isolated from the infected experimental animal.
6. A specific immune response to the virus can be detected.

None of these four studies met all six postulates. Of the four studies said to prove that a coronavirus causes this disease, not one of them satisfied the first three postulates, and none of them even addressed postulates four and five. One paper claimed to find an immune response (postulate six) by looking at antibody levels from the patient.

The first two papers are honest enough to claim only an association of coronavirus and the disease; the third paper claims that the coronavirus is "identified as the causative agent." The fourth paper, from McMaster University, falsely claims that the coronavirus *is* the causative agent of the disease and that the virus "set in motion the pandemic," with no evidence to back up these statements.

These papers never show that all the people with Covid-19 had the same set of symptoms; they never purify any virus from the sick people; they never demonstrate the absence of the virus from well people; and they never show that the transmission of purified virus could make well people become sick. This is scientific fraud of the first order.

It's interesting to look more closely at how virologists work to "prove" something like causation by coronavirus. One example is a paper published in 2003 in *Nature*, titled "Koch's Postulates Fulfilled for SARS Virus."[11] Researchers claimed that Severe Acute Respiratory Syndrome is caused by a coronavirus. The title itself is misleading, if not dishonest, because the researchers satisfied neither Koch's nor Rivers' postulates.

Here's what they did: first they took respiratory secretions from some sick people; in other words, they took sputum from people with a cough. They centrifuged the sputum, which separates the cellular part (where presumably the virus is residing in the cells) from the liquid part. They discarded the liquid part. This is what they referred to as "purification." Then they took this centrifuged, unpurified sediment from sick people, containing God-only-knows-what, and inoculated that into vero (monkey kidney) cells. Here we have to understand that if virologists want to

get enough "virus" to use experimentally, they must grow it in a biological medium such as an animal or at least cells from an animal. Unlike
bacteria, which can be grown in petri dishes, viruses are not alive, and
they can "grow" only in other living cells. For convenience and because
cancer cell lines are "immortal," they generally grow their "viruses" in
cancer cells; however, in this case they used kidney cells. This practice
is fraught with obvious problems for proving it is the virus and not the
kidney or cancer cells that are causing disease when these viruses then
get injected into the test animals. Also, it is well known now that as part
of their "detoxification" strategy, cells, especially cancer cells produce
particles called *exosomes*, which are *identical* to "viruses." (More on this
in chapter 6.)

Again, the researchers took unpurified sediment from the nasal
mucus of sick people and grew that in vero cells until they got a sufficient
quantity of cellular material to work with. Then they centrifuged this
mess again, not even attempting to purify any virus from the mixture.
Finally, they took this witch's brew of snot sediment, kidney cells, and
who-knows-what-else and injected that into two monkeys. They didn't
do a control group by injecting saline into other monkeys or injecting
vero cells into monkeys, or even injecting the liquid supernatant from
the centrifuged material into monkeys. They just injected this cellular-
debris-laden goop. One monkey developed pneumonia and the other
appeared to have respiratory symptoms possibly related to a lower respiratory disease. That, claim the researchers, is the proof that a "coronavirus"
can cause disease.

To be fair, in a related study,[12] researchers did the exact same procedure, except to make it more reflective of how the new "virus" actually
spreads, they took unpurified, lung-cancer-grown, centrifuged snot and
(again, without any controls) squirted it down the throats and into the
lungs of hamsters. (Where is PETA when you need them?) Some, but
not all, of the hamsters got pneumonia, and some died. We have no idea
what would have happened if they had squirted plain lung cancer cells
into the lungs of these hamsters, but probably not anything good. And
even more perplexing, some of the hamsters didn't even get sick at all,
which certainly doesn't square with the deadly, contagious virus theory.

In short, no study has proven that coronavirus, or indeed any virus
is contagious, nor has any study proven anything except that virologists
are a dangerous, misguided group of people and that hamster- and mon-
key-rights people are not doing their jobs!

This story is analogous to "proving" that the ping-pong ball can knock down walls by throwing a bucket of rocks and ice cubes containing a single ball at a small wall and showing that this does knock down the wall. These "proofs" make no sense and prove nothing, and yet the whole edifice of corona "virus" causation rests on these bogus studies. In chapter 5, we will deconstruct the equally bogus "testing" now used to provide what passes for supportive evidence of a viral causation. Stay with us, the ride gets more interesting as we go.

CHAPTER 5

TESTING SCAM

In a series of recent articles published in local and national news media, as well as in various scientific papers, some of the world's leading medical doctors and immunologists have made surprisingly honest—and shocking—statements about coronavirus testing. The test used is called a PCR test—PCR means polymerase chain reaction. Here are some samples of these statements:

"We did not perform tests for detecting infectious virus in the blood."[1] This statement came from a paper in which the authors allege they had discovered the novel coronavirus in patients suffering from Covid-19.

"There is no way to tell whether the RNA being used in the new coronavirus PCR test is found in these particles seen under the electron microscope. There is no connection between the test and the particles, and no proof that the particles are viral."[2] Ironically, this statement does not come from a paper attempting to debunk the coronavirus causation of Covid-19. It comes from a paper that adamantly espouses this connection, but which claims that more research is needed to understand the interesting intricacies of this new virus.

Or this, from the head of the Marin County Public Health Department, charged with making public health policy for Marin County, California: "The PCR tests means you either are infected with the novel coronavirus or you are not infected with the virus."[3] Such a statement is akin to going to the refrigerator store to purchase a new refrigerator and asking the salesman about a new model in the showroom. He says to you, "It's a new and interesting model; it will either keep the food cool or it won't." Most people would not choose to elevate this person to head of refrigeration for Marin County.

An NBC news feed reported on certain puzzling results: sailors who test positive, then negative, then positive for Covid-19 using the PCR test. This should not happen if the test is any good. A CDC representative stated, "Detection of viral RNA does not necessarily mean that infectious virus is present."[4]

According to one infectious disease expert at Vanderbilt University Medical Center in Nashville, "It's possible that people could shed remnants of the virus for some period of time. That doesn't mean anything is wrong with them or that they are contagious."[5]

These statements are all referring to the test used to claim that a person is infected and can spread the disease! When asked about these statements, the head of a lab that specializes in testing for infectious disease told us: "A positive PCR test means you have active disease or you are a carrier and don't have active disease." When we asked how to distinguish between those who have active disease and those who are carriers, she confidently replied: "Those with active disease are sick, the carriers are well."

Then we asked how one knows whether sickness is caused by the virus, since the test can be positive whether or not you are sick. She replied, "You can do a PCR test and find out whether the sick person has the virus." Welcome to the Alice-in-Wonderland world of virology!

Finally, the most revealing quote of all, this one from the chief of infectious diseases at Wake Forest Baptist Health in Winston-Salem, North Carolina: "We just don't have enough details yet to make confident statements about immunology."[6] This quote is by an immunologist, and immunologists are the ones deciding public policy. They have put the world under house arrest. It would be nice if they could at least confidently say they knew something about immunology.

How did this shocking situation concerning tests for viral diseases come about? Let's return to the story of Stefan Lanka, PhD, a virologist from Germany, whom we discussed in chapter 3. Lanka's work has helped cut through the veils behind which the field of virology is shrouded. As a young graduate student in Germany, Lanka made the chance discovery of the first virus in seawater. Using electron microscopy in his studies of sea algae, he noticed that the algae contained "particles." To find out what these particles were, and knowing that no one had described viruses living within healthy algae before, he proceeded as follows: He ground up the algae in a sort of blender, essentially to break apart the walls of the algae. Then he purified this mixture using an extremely fine filter to separate out particles the size of viruses from everything else. In this way, he obtained a pure solution of water and viruses and anything else that is the size of a virus or smaller. Then he put this mixture into a density-gradient centrifuge, which spins the solution and allows the particles to separate out into bands. The final step uses a micropipette to suck out the band that contains *only* the virus.[7] This simple procedure is the gold standard for the purification and isolation of a virus from any tissue or solution. It's not an easy process, but it is not unduly difficult either.

He could then study this purified virus under an electron microscope, elucidate its shape and structure, analyze the genome, and ascertain which proteins it contained. With this work, he could confidently state he had discovered a new virus and was sure of its makeup. For this discovery, he received his doctorate and was about to embark on a promising career as a virologist.

The only part of Lanka's experiment that surprised him was that in studying the interaction of algae with this new virus, he was forced to come to the conclusion that algae containing the virus were thriving and were much healthier than the algae without the virus, which were barely surviving. He may have been the first to come to the conclusion that real viruses in the bodies of other species are not pathogens (as was thought at the time) but rather are integral to the healthy functioning of the host. In essence, he was one of the first to propose that in addition to having a microbiome inside us, we also have a *virome*; and without this virome, we cannot be healthy. This was a radical concept in the 1980s, as no one else had proposed such a theory.

If we compare the simple, logical, straightforward way in which Lanka isolated, purified, and characterized his virus, with the description of how modern virologists propagate viruses now, one begins to see the problem and confusion around testing for viral diseases. When Lanka realized that workers in the field of modern virology were never isolating, purifying, or properly characterizing "viruses" but instead confusing what they were finding with artifacts made by their propagation techniques, he naturally questioned whether the viruses that supposedly were causing disease even existed. His question was not so much about whether viruses are infectious entities, but something even more fundamental—whether these viruses even exist at all.

Compare Lanka's careful work with how current virologists find and characterize viruses, including the "novel coronavirus." They start with sputum from a sick person, having no idea how this person got sick. They centrifuge but do not filter the sputum. This is not a purification process, as they readily admit in all of the papers written about the "coronavirus."

Here's what the authors of the original papers that found and linked the "novel" coronavirus (SARS-CoV-2) to the disease now called "Covid-19" have to say. The following quotations come from the brilliant paper "Covid-19 PCR Tests are Scientifically Meaningless," by Torsten Englebrecht and Konstantin Demeter.[8]

Referring to a published image in a paper claiming to have isolated a new virus, they say, "The image is the virus budding from an infected cell. It is not purified virus."

If it's not purified virus, how do the authors know whether or not it is even a virus, what it is, or where it came from?[9]

In the paper "Identification of Coronavirus Isolated from a Patient in Korea with Covid-19," the authors stated: "We could not estimate the degree of purification because we do not purify and concentrate the virus cultured in cells." In other words, they did not isolate the virus, even though they claim to do so in the title.[10]

In the article "Virus Isolated from the First Patient with SARS-CoV-2 in Korea," the authors admitted, "We did not obtain an electron micrograph showing the degree of purification."[11] In other words, the authors have no idea whether or not the sample is purified, as an electron micrograph is the *only* way to determine that. They then claim to have characterized the genetic material of something they never purified, having no idea what they were looking at. This was an important study, as it describes the first case of "Covid-19" in Korea.

Finally, the article "A Novel Coronavirus from Patients with Pneumonia in China," states: "We show an image of sedimented virus particles, not purified ones."[12] The researchers took nasal mucus ("snot") from sick people, centrifuged it (which is *not* a purification step) and then showed a fuzzy picture of what they found. Then they carried out a "genetic analysis" of this sediment in order to characterize the "novel" coronavirus. This piece of fraud was published in the esteemed *New England Journal of Medicine.*

What is in the centrifuged material that these papers describe? The centrifuged material contains bacteria and perhaps viruses, fungi, human cells, cell debris, and anything else found in the lungs or sinus passages of a sick person. Researchers then inoculate this unpurified mess onto "living tissue" to make it "grow." Sometimes this tissue is lung cancer tissue, sometimes aborted fetal tissue, and sometimes tissue from monkey kidneys. In any case, it is a complex mixture of many components, known and unknown. And then because this "virulent, infectious virus" won't infect and kill this living tissue unless you starve and poison the tissue first, you deprive the tissue of nutrients and add oxidizing agents to "weaken" the tissue. Then you add antibiotics to make sure it's not bacteria that is killing the tissue.

The tissue from this treatment naturally disintegrates into thousands of components. Then you centrifuge this mess again to find your "virus." At that point, you start the PCR testing to determine the genetic and protein makeup of this "virus." The problem is that (unlike the clear situation that Lanka encountered) in this slapdash way, you *never* have the isolated intact "virus" as a reference to allow you to know which genetic parts of your unpurified mess actually belong to the "virus" you are trying to characterize.

As mentioned in chapter 3, after studying the way in which virologists said they had found the measles virus—without isolating and purifying it, and actually deciding on the genetic makeup by consensus—Lanka offered a prize of one hundred thousand Euros to anyone who could demonstrate its existence. In the first court to hear the proceedings, the claimant for the prize won the case, the judge concluding that proof of the measles virus did indeed exist. However, the German Supreme Court, with its more stringent rules of evidence and the appointment of a science master to oversee the case, ruled that the claimant, in fact, did *not* prove the existence of the measles virus. Lanka did not have to pay the claim.

How is Lanka's work relevant to the current test used to detect the presence of viruses, specifically for the coronavirus? Clearly, if one can't prove that the coronavirus even exists and that the testing for this imaginary virus is bogus, then the world has been led wildly astray. If the test for the coronavirus is inaccurate and misleading, as is the case, then there are no grounds for believing the reports about the number of Covid-19 cases, the number of Covid-19 deaths, or any other statistics coming from the orthodox medical institutions. If the testing is bogus, then the coronavirus emperor has no clothes.

Let's contrast Lanka's elegant experiments with the testing procedures used to determine the alleged presence of coronavirus (SARS-CoV-2) "infection."

The first thing we need to understand about a PCR (polymerase chain reaction) test is that it is a surrogate test—it does not find a virus; rather, it finds something else said to indicate the presence of the virus. A surrogate test is one that is generally easier and less expensive to perform and can stand in for the gold-standard test (of actually finding the virus) and thereby make the clinical practice of medicine easier, safer, and cheaper.

For example, pulmonary emboli result from blood clots that travel to the lungs. Symptoms include chest pain and shortness of breath. Pulmonary emboli can be fatal. It is important to diagnose the condition early, as it can be treated with blood thinners. It is also important to diagnose accurately, as pulmonary emboli share many symptoms with heart attacks and pneumonia, which call for different kinds of treatment.

Fortunately, there is a test that is 100 percent reliable for finding pulmonary emboli when preformed properly. The procedure, called an angiogram, involves inserting a catheter into the arteries of the lungs. Then the radiologist injects dye into the artery; the dye contains heavy metals that one can see on an X-ray. If a clot is present, the angiogram reliably and accurately demonstrates its presence to the radiologist on the real-time X-ray. With this test, referred to as the "gold standard," one can confidently say whether or not the patient has an embolus.

Angiography, however, is technically challenging. It is difficult to find the artery with the catheter. It is costly, due to the time and equipment needed. It is dangerous, as the artery may tear during insertion of the catheter. Another problem with angiography is that it requires injecting heavy metals into the artery and subjecting the patient to a lot of radiation.

Consequently, medicine has looked for a surrogate test that can more easily and safely detect pulmonary emboli. The V/Q scan looks at the blood flow into and through the lungs and compares this with the air movement into and through the lungs. When all is well with the lungs, these two parameters match up. When an embolus is present, they usually don't match up because blood flow is compromised. This allows the diagnostician to surmise that even though they have not actually seen a clot, it is likely to be present.

A *surrogate* test is one that does not look for what you need to find but rather for something likely to be there if the condition is present. The surrogate test allows doctors to make an educated guess. To validate a surrogate test, one must first do a careful study in which the surrogate test is compared to the gold standard test. This gives you exact information as to how accurate you can expect the surrogate testing to be. These validating studies are usually carried out at a large medical center or a group of medical centers. The researchers begin by finding a large number of patients—say two thousand—with symptoms typical of pulmonary emboli. For simplicity's sake, let's say that one thousand do have an embolus as demonstrated with an angiogram, the gold standard test, whereas the other one thousand do not. Now, we have a group of people who either have or don't have the condition we are testing for. Then we do a V/Q scan on each of these two thousand patients. In the group that we know has an embolus, if nine hundred are positive on the V/Q scan, then we know that the surrogate test picks up the diagnosis 90 percent of the time. In the other 10 percent, for whatever reason, the surrogate test fails to demonstrate the emboli even though we know it's there. We now know that the false negative rate is 10 percent. This allows the clinician to forgo the more difficult angiogram because they know how likely the V/Q scan will be to pick up the embolus if it is there. They also know that in the event that the test is negative they still have a 10 percent chance the test missed it. In that case, they may want to move to the angiogram if the level of suspicion is high that the patient does, in fact, have an embolus in spite of the negative V/Q scan. This is basically the art of modern medicine.

Likewise, the researchers can then take the one thousand people who test negative on the angiogram, do a V/Q scan on all of them, and determine the false positive rate. If one hundred subjects test positive on the V/Q scan even though you know for sure they have no clot, then you can accurately and confidently say that for whatever reason 10 percent of the time the V/Q scan is saying you have a clot when you don't. Again, this helps the clinician who may be confronted with a patient whom he is pretty sure doesn't have a clot (for example, they may have evidence of pneumonia or a heart attack) but orders a V/Q scan anyway. If the V/Q scan is positive they may want to confirm this by moving on to an angiogram because they know that in 10 percent of the cases the V/Q scan is falsely positive. Clearly the lower the number of false positives and false negatives, the better and more reliable is the test.

The point is, without having a gold standard test with which to compare your surrogate test, and without this comparison having been performed in the clearest and most meticulous way possible, there is *no possibility* of having an accurate surrogate test. To be even clearer, without this comparison, the surrogate test is completely useless and misleading . . . completely—yet officials are using surrogate Covid-19 tests to send people into nursing homes, remove children from their families,[13] and even separate newborn babies from their mothers if the mom tests positive![14]

PCR tests, antibody tests, and every other test for a "coronavirus" are surrogate tests, which have never been compared to any gold standard; therefore, they are completely and utterly useless and misleading. They are propaganda, not science.

The gold standard test for a viral infection is the isolation, purification, and characterization of the virus (as outlined in the description of Lanka's experiment) and the proving of contagion. Lanka didn't prove that the virus he found was contagious simply because it isn't, and no virus is contagious. This is the only possible gold standard there can be.

The PCR test examines the pieces of genetic material taken from a swab from the back of the sinus cavity (a very unpleasant procedure). No research has shown that this genetic material is unique to that of coronavirus or even that it comes from a coronavirus.

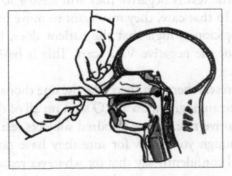

Taking a swab for a PCR test—the procedure is invasive and unpleasant, and for some it is even painful.

Furthermore, in order to examine this genetic material, the test "amplifies" it. An amplification cycle means they start with a probe that matches the snippet of DNA or RNA they are looking for. Then, because this is too small to detect, they repeatedly double the snippets. If the sample changes the color of a solution, the test is considered positive. If you do too few amplification cycles, you never find the snippet, resulting in a false negative. If you do too many amplification cycles, you find the snippet too often because the test also amplifies the background genetic snippets ("noise"). These are false positives.

One can therefore manipulate the amplification cycles to get whatever result one wants. Too few cycles and everyone tests negative; too many cycles and most test positive.

John Magufuli, president of Tanzania, may be the wisest world ruler alive today. A chemist by training, Magufuli submitted samples to the World Health Organization (WHO) for testing. Said Magufuli, "We took samples from goats; we sent samples from sheeps; we took samples from pawpaws; we sent samples from car oil; and we took samples from other different things; and we took the samples to the laboratory without them knowing." His officials named the sample of car oil Jabil Hamza, thirty years old, male. The results came back negative. They named a sample of jackfruit Sarah Samuel, age forty-five, female. The results came back inconclusive. Pawpaw got sent in as Elizabeth Ane, twenty-six years old, female. The poor pawpaw came back positive. Samples from a bird called *kware* and from a goat also tested positive; rabbit was indeterminate; sheep was negative.[15] President Magufuli is not wasting any government money on testing for his people, but in the West, governments have spent millions for the PCR test kits.

Since no PCR test has ever undergone comparison to any gold standard, the results are meaningless. This is not a situation where we just need better or more accurate testing. As Kary Mullis, the inventor of the PCR technology, has insisted, time and again, PCR tests do not prove causation and cannot diagnose illness.[16]

The CDC and the FDA concede that the PCR test cannot be used for diagnosis. A file from March 30, 2020, stated: "Detection of viral RNA may not indicate the presence of infectious virus or that 2019-nCoV is the causative agent for clinical symptoms" and "This test cannot rule out diseases caused by other bacterial or viral pathogens."[17]

Furthermore, the FDA admits that "positive results ... do not rule out bacterial infection or co-infection with other viruses. The agent detected may not be the definite cause of disease."[18]

According to a product announcement for the LightMix® Modular SARS-CoV Assays, "These assays are not intended for use as an aid in the diagnosis of coronavirus infection."[19]

One can only wonder what exactly the test is supposed to do if not diagnose a coronavirus infection?

The same methodological problems are found with the new antibody tests for assessing immunity to "coronavirus." Antibody testing is
another type of surrogate testing that does not diagnose illness or ascertain its cause. A brilliant paper by David Crowe[20] explains in detail the
fact that the theoretical underpinnings of antibody testing have not been
demonstrated in any experiment. This is why the Wake Forest immunologist had to admit, "we don't know enough about immunology to make
any conclusions."

Scientists believe that antibodies have a predictable and exact course
as they track a viral infection. Antibodies are proteins made by our
immune system to either fight off an illness or "remember" that we have
encountered an infectious organism like a virus—at least, this is what
we've been told. The theory is that before we encounter a virus or get
sick from a virus, we have no antibodies to it. After getting sick, the
PCR test should detect the virus (or, more accurately, the genetic pieces
we think could have come only from that virus). Then after one week
(because viruses and our immune system seem to understand the concept of one week), we begin to make an antibody called IgM, which is
not specific to this coronavirus but is made by our immune system to
fight off any virus. Then on day fourteen, as the virus is cleared from
your body, the PCR test goes back to negative, the IgM levels go down,
and we see the appearance of the more targeted IgG antibody. Then at
day twenty-one (because viruses understand that this happens in weekly
intervals), the IgM is gone, the antibody test is reliably negative, and
the IgG has peaked. At day twenty-eight (because your immune system
also understands weeks), the IgG level drops to a level it can sustain for
the long term. Once the test manufacturer or virologist sees a stable IgG
level, they supposedly know you are immune for life from the effects of
the virus . . . or maybe not.

This theoretical and imaginary construct has many holes big
enough that you could drive a truck through them. Consider the following: it turns out that a small percentage actually have IgM, IgG, or
both antibodies months before having the "infection."[21] How this is
possible, since this is a novel virus that humans have never seen before,
is not explained.

Finally, studies show that IgG sometimes appears before and sometimes appears after IgM; sometimes there is no IgM; sometimes there
is no IgG.[22] In either case, it may mean you either had the virus or you

didn't. And as with AIDS, there is no proof that a particular level of IgG confers immunity.

Ah, but virologists have an explanation for this: these new wily viruses somehow know how to escape detection and neutralization even if the person has a robust antibody response! That, of course, is nonsense.

Then we're told that a positive PCR test means you are either sick or not sick, infectious or not infectious, and that sometimes the test is positive, then negative, then positive, then negative. It's enough to make even the Mad Hatter go silent with disbelief!

CHAPTER 6

EXOSOMES

After reading the last two chapters you may be shaking your head in disbelief; you may have so many questions swirling in your mind that you feel disoriented. The main question for us all is how the entire world of medicine, virology, and immunology, along with our political leaders, could have made such an obvious mistake? How could generations of doctors and researchers have become convinced that many of our common diseases are viral in origin?

Let's first provide the scientific basis for challenging the notion of *contagion*. As we have said, an in-depth look at the scientific literature reveals no proof of the contagion theory, but the alternate explanations for so-called "bacterial" or "viral" illness do have research behind them. Only Western medicine invokes the concept of contagion—person-to-person transmission of harmful bacteria or viruses. Neither traditional Chinese medicine (TCM) nor Ayurveda (a system of medicine with historical roots in the Indian subcontinent) entertains the concept of contagion. These ancient healing systems look at imbalances, diet and toxins as the causes of disease.

So how did the theory of viral causation come about? During the late 1800s, with the popularity of Pasteur and the increasing materialistic thinking of the age, the germ theory gained popularity. The germ theory explained common observations, such as why drinking sewage water made people sick and why people who share a workspace or household seem to get sick in a similar way at the same time. With the advent and widespread use of the light microscope, scientists and doctors could clearly identify bacteria associated with particular illnesses.

In the nineteenth century, scientists and physicians assumed that the teeming forms they saw in their microscopes caused disease and were hostile to life. In writing *On the Origin of Species* (published in 1859), Charles Darwin (a contemporary of Pasteur) proposed a theory of evolution in which only the plants and animals best adapted to their environment survive to reproduce. He painted a picture of life in which the various organisms were in constant struggle against each other. Darwin borrowed popular concepts (such as "survival of the fittest") from sociologist Herbert Spencer and "struggle for existence" from economist Thomas Malthus. The notion of hostility and competition in all of nature fit with attempts to justify the social inequalities, poverty, and sufferings that characterized the dawning Industrial Age. Social Darwinism actually preceded biological Darwinism!

For all the known "infectious" bacterial illnesses, the science points to other accurate explanations—namely starvation and poisoning. However, the microscope gave scientists the ability to find germs at the site of disease. Their observations revolutionized the practice of medicine and our thinking. The microscope allowed medicine to enter a "scientific" age and provide a ready and easy explanation for illness—one that circumvented the more difficult and less profitable work of cleaning up the cities, improving diets, mitigating poverty, and reducing pollution.

However, bacteria are found at the site of disease for the same reason that firemen are found at the site of fires. Bacteria are the cleanup crew tasked with digesting and getting rid of dead and diseased tissues. Claiming that bacteria cause a certain disease is no more reasonable than claiming that firemen cause fires, especially as experimental evidence shows this to be false. Likewise, maggots on a dead dog are there to clean up dead tissue—no one would accuse the maggots of killing the dog. In fact, one therapy for necrotic tissue is maggot therapy (applying maggots to the wound). The maggots eat only the dead tissue; when there is only live tissue left to eat, they die off.

But scientists could not always find an offending bacterium for a specific disease. Louis Pasteur could not find a bacterial agent for rabies, and he speculated about a pathogen too small to be detected using a microscope.[1] The same held true for polio—hard as they tried, scientists could find no bacteria at the site of the illness.[2] Following Pasteur, and completely wedded to germ theory, they postulated a tiny enemy, something that our technology could not yet visualize. The search was on to find this disease-causing organism.

The eureka moment came with the invention of the electron microscope; scientists finally saw tiny "particles" at the site of disease. These particles had "stuff" inside them, suggesting they were "alive." They were more abundant in diseased tissue than in healthy tissue (although this is not what Lanka found in algae). There were variations between types of particles, suggesting that one type of particle caused one disease and another particle type caused a different disease. Immediately assuming that these particles were bad for us, scientists named them *viruses*, after the Latin word for "toxin."

Further research revealed that these particles often emerged from within the cell; this led to the conclusion that these viruses were not just bad for the cell in which they resided, but they could invade other cells. Scientists surmised that viruses co-opted the "machinery" of the cell like parasites, turning the cells into "slaves," meaning the cell would do the bidding of its new master, the infecting particle. Like alien invaders in science fiction movies, the particle would come from the outside, inject itself into the cell, take over the genetic machinery of the cell, reproduce itself by the thousands, and then emerge from the cell to continue on its evolutionary path, spreading to take over the world.

The wily virus theory was born—except that what scientists had really discovered with their electron microscopes was not viruses but exosomes. The only thing infectious in this scenario was the noxious belief that these small particles, dubbed viruses, caused disease. This false theory was the part that spread all over the world and is now threatening to kill us all.

Exosomes are simple, well-characterized features in the cells of all creatures, and conventional scientists have carefully elucidated their functions.[3] When a living organism is threatened in almost any way—through starvation, chemical poisoning, or electromagnetic effects—the cells and tissues have a mechanism for "packaging," "propagating" and releasing these poisons. Modern researchers have shown that exosomes have exactly the same attributes as "viruses." They are the same size, contain the same components, and act on the same receptors.[4]

HIV researcher James Hildreth, president and CEO of Meharry Medical College and former professor at Johns Hopkins, put it this way: "The virus is fully an exosome in every sense of the word."[5] Exosomes are completely indistinguishable from what the virologists have been calling "viruses."

Here's how exosomes work: let's say you have a poorly nourished organism, then you expose it to a common environmental toxin. The tissues and cells that are affected begin to produce, package, and secrete these poisons in the form of exosomes. This is a way of ridding the cells and tissues of substances that would do it great harm. The greater the exposure to toxic assaults, the more exosomes will be produced.

Studies have shown that if one somehow stops the cells from producing and excreting these exosomes, then the cells and tissues, in fact the organism, will have a worse outcome.[6] This research demonstrates that the production and excretion of exosomes is a *crucial* detoxification function of all cells and tissues.

Another clearly demonstrated function of these exosomes is that they act as a kind of key that circulates in the blood and lymph of organisms, such as mammals and humans, until they find a distant cell with a lock into which this key fits.[7] The exosome acts as a messenger, essentially warning the other cells and tissues that danger is afoot and that they need to prepare.

Exosomes leaving a cell.

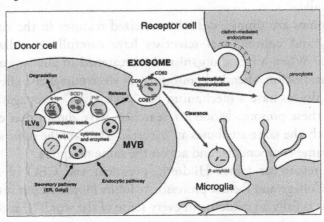

Exosomes carry messages from one cell to another.

Far from acting as hostile invading viruses, exosomes provide a brilliant system of communication within an organism to rid the cells and tissues of poisons and then communicate to the rest of the organism what has happened.[8] Far from acting as a source of illness, these particles are an integral part of our detoxification system. They are the true firemen, obviously present in higher amounts in cases of disease, in which a higher burden of poisoning has occurred.

In fact, these "viruses" are not invaders but toxin-gobbling messengers that our cells produce to help us adjust to environmental assaults, including electro-smog.[9] After all, most people have adjusted to worldwide radio waves, electricity in our homes, and ubiquitous Wi-Fi—and the sparrow population rebounded after the Great Plague of 1738. It is exosomes that allow this to happen. These tiny messengers provide real-time and rapid genetic adaptation to environmental changes. (Whether these exosomes can help us adapt to the extreme disruption of 5G is the question of the day.)

If you do an Internet search, you will find that exosomes are the latest thing in medicine, used as a treatment for cancer, antiaging products, facial rejuvenation, hair regrowth, and even "exosome penis treatment."[10]

Finally, the research shows that toxic exposure, including exposure to fear and stress, increases exosome production.[11] This should come as no surprise to any honest observer of sickness and health since many have noted that stressed, worried, fearful people get sick easier, so it makes sense that you would find increased detoxification "products" in their tissues.

There is now clear experimental evidence that exosomes made by one organism can be picked up by other organisms (of the same or different species) and cause protective reactions in these new organisms.[12] One study showed that if mice are exposed to the liver toxin known as acetaminophen (Tylenol), the liver cells increase their production of protective exosomes. The researchers isolated and purified these exosomes and exposed other mice to them. The second group of mice did not get sick, as would be predicted by the virus theory; instead they developed protective responses in their livers and secreted more exosomes.[13]

This is similar to what trees do when confronted with a beetle infestation. The originally affected tree produces chemicals that help the tree survive the beetle exposure. These same chemicals are secreted, with the help of the fungus or mycelium in the soil, through the root system of the tree. These chemicals then serve as messengers to the surrounding trees, telling them that beetles have taken hold and that protective measures may be needed. If the beetles go away, then these measures are not taken; if the beetles show up, the surrounding trees also produce a protective response.

The real point here is that thanks to exosomes, nature is not raw in tooth and claw but a superb cooperative venture. The originally affected tree is not competing for survival with the other trees; the affected tree needs the other trees in order to survive and thrive. We need each other—members of our own species and other species—otherwise none of us will survive.

The germ theory is wrong; the virus theory is wrong. Viruses are not here to kill us; in reality they are exosomes whose role is to provide the detoxification package and the communication system that allows us to live a full and healthy existence. A war on viruses is a war on life.

It's clear that the misidentification of exosomes as viruses was a tragic mistake, one that it's about time we correct, once and for all.

What we know about exosomes can help us solve the mystery of childhood diseases like chicken pox and measles, and also of STDs, which seem to require an explanation of "contagious." This will be the subject of the next chapter.

CHAPTER 7

RESONANCE

Recently, on a forum for scientists, laypeople, and health-care professionals who question the safety and efficacy of vaccinations—in other words a forum for people who question the practices of conventional medicine—a professional scientist made the following comment when one of the members pointed out that Koch's postulates had never been met for even a single viral or bacterial disease. She said, "No virologist of any note took any notice of Koch's postulates ever."

The comment says a lot about how virologists currently think, as Koch's postulates are what anyone with common sense would use to prove that a microorganism causes disease—isolate the organism from a sick animal or human and then introduce it into a healthy animal or human to see whether it causes disease. In a sane world, Koch's postulates are not something one can "discard."

In the mid-twentieth century, virologists apparently came to a fork in the road. With repeated failures to satisfy Koch's or Rivers' postulates, it became obvious that viruses don't cause disease. The scientists could either admit this and all become postmen, butchers, and greengrocers—in other words get honest jobs—or they could proclaim that they changed the rules of logic, hope no one would notice, and possibly get fabulously wealthy and powerful from patents on antiviral medications and vaccines. It's actually understandable that they took a flyer and chose the second route. Tragically, this farce worked, and the world became a much worse place for living beings to inhabit.

The scientist quoted above later said:

Polio alone revealed to them that out of every thousand people infected, around ten would show signs of illness and one would become paralyzed. Even the old works on neisseria [a type of bacteria said to cause meningitis and syphilis] from the old days, showed conclusively that the bacteria was carried extensively and routinely and that only around one in one hundred thousand would actually have clinical meningococcal disease presentation. So to toss in Koch's postulate in order to dismiss an article is chasing an unrelated goose down another garden path.

Suppose a professor theorizes that spraying a gentle stream of water (such as a shower) on people would kill them. To test the theory, one hundred people are sprayed. No one is harmed. Most people would conclude the theory was incorrect and spraying water on people, in fact, doesn't kill anyone. But some researchers would persist. They would test a thousand people, then fifty thousand. . . . Finally, by chance, one subject dies. Of course, no honest or sane person would keep doing this experiment this long, but if it were done, the obvious question to ask would be whether something else happened to the person to kill him as it obviously couldn't have been the water. Maybe he slipped in the shower, hit his head, and died; or maybe he had an aneurysm in his brain, which due to a fight with his wife right before being sprayed, happened to burst; or maybe he foolishly decided to spray the water into his airway to wash his dirty lungs. In any case, the cause of his demise is clearly *not* the spraying of water and the researchers would need to do an individual investigation to find out what had really happened.

For a prominent alternative leader not to understand the point of Koch's postulates and write this in a public forum is cause for despair. One can only conclude that the depth of the delusion in the biological sciences is so deep that even scientists supposedly dedicating their lives to uncovering one aspect of the delusion can't step out of the whole delusion and see things clearly.

Another online statement by a prominent scientist goes even further:

Koch's postulate is completely wrong and irrelevant in our modern understanding of disease-causing agents. That is a set of principles developed in 1884! It was ten years before viruses were even discovered, and over sixty-five years before the immune system was discovered and of course one hundred fifteen years before the microbiome was understood. Most of the tenets of Koch's postulates are wrong. Many, many well known infectious

agents don't fit it. If any doctor or scientist is out there using Koch's postulate as "proof" that this Cov2 is not real, turn away because they have no
idea what they are talking about. Just like we moved away from the early
1600s geocentric model of the solar system, we have to move away from
Koch's postulates.

This is like saying that because Newton formulated the laws of gravity over three hundred years ago, they are now out of date and it's safe to
jump off tall buildings!

A remaining question is how to explain "measles parties" and sexually transmitted diseases (STDs) like herpes. Investigating these phenomena leads to some interesting conclusions about the nature of life.

To understand what seems to be the contagious nature of childhood
diseases like measles, mumps, and chicken pox, or STDs like herpes,
gonorrhea, or syphilis, one must investigate the phenomena of *resonance*.
If one plucks a string tuned to a certain frequency, the vibrations of the
string will cause a second string tuned to the same frequency to vibrate
and sound at the same frequency. The two strings are not touching; the
connection is through a sound wave that travels between the strings.

When confronted with the question of "what is a virus?" one could
easily ask that question about anything in nature. We could say that a
virus is made of chemicals—proteins, nucleic acids, minerals, lipids, etc.
But what are these chemicals made of? They are made of atoms, such as
sulfur, oxygen, and carbon. Atoms are made of protons and neutrons in a
nucleus and electrons circling around the nucleus, sort of like the sun with
the planets circling around it. And as in the solar system, 99.999 percent of
this atom is space; that is, it seems to be nothing.

This presents an obvious dilemma. How is it possible that this particle,
which is made almost entirely of nothing, creates an entity we call a virus,
or a foot or a chair, all of which seem solid? The only conclusion anyone of
integrity can make is that we simply have no idea how this works. To make
matters even worse, the top dogs of the science world, the physicists, tell us
that all this stuff on earth can exist either as a space-filling particle or as a
wave, which has no physical presence at all. In other words, both humans
and viruses are made of waves of energy that have no discernible physical
presence. How a collection of waves can write a book on viruses is anyone's
guess, but the fact that this occurs can't be disputed.

The only rational conclusion anyone can come to is that physical
reality is a kind of energy or wave pattern that crystallizes as physical

reality under certain conditions. We and everything in the universe seem to participate in this dance of wave-particle manifestation.

With this understanding, let's turn to the results of a series of experiments carried out by a virologist named Luc Montagnier. Montagnier is credited with claiming to discover that the HIV virus causes AIDS. (He also claims that the Covid-19 "virus" is man-made.[1]) What he found can help us craft a realistic theory explaining the mystery of childhood illnesses like measles and STDs, which appear to be contagious (and also proves that even the misguided can sometimes redeem themselves).

We need to be cautious about applying his findings too broadly. The vast majority of illness that seems to be contagious is in reality just people exposed to similar toxins or suffering from the same nutritional deficiencies. Hiroshima was not contagious; Chernobyl spread throughout Europe, but it was neither contagious nor caused by a virus. Sailors all getting sick on the same ship were not the victims of a virus; the likely explanation is that they all had a vitamin C deficiency called scurvy. Young people at college exposed to horrible food, severe psychological stress, and intense binge alcohol use are affected by toxins, not some elusive virus.

After starvation and toxicity are accounted for, we can admit that some diseases can be spread by a kind of energetic resonance as predicted by a careful and accurate study of the nature of physical stuff, carried out by Luc Montagnier.

Here's how the experiment goes: first, one puts DNA or RNA in water (beaker one). Then one puts a collection of nucleic acids (the chemicals that make up the DNA and RNA) in a separate water beaker (beaker two), in another part of the room. Then one introduces an energy source, such as UV or infrared light and shines that on beaker one, which contains the formed DNA or RNA. In time, the exact same sequence of DNA or RNA will form out of the raw materials in beaker two.

There is no possibility of physical connection between the two beakers. The only conclusion one can draw from this simple experiment is that the DNA or RNA in the first beaker has a resonance energy picked up by the second beaker. This resonance energy then becomes the blueprint for the formation of the identical piece of DNA or RNA in the second beaker.[2] This revolutionary experiment is clear and simple—and repeatable.

This formation of DNA or RNA in the second beaker can happen *only* if both beakers have water in them. Without water, no resonance is possible. Even in our string example, it is the water vapor in the air that is resonating.

When one applies this discovery to viruses (or exosomes) said to cause measles, chicken pox, or herpes, it is possible that since these particles called viruses or exosomes are simply packages of DNA or RNA, they emit their own resonant frequencies. In a way not yet determined, each frequency creates an expression that we call a disease; however, the frequency will create what we call illness *only* if there is a purpose or reason for the illness.

Chicken pox is a universal way for children to live a long life. Children who experience chicken pox have less disease (and especially less cancer) than do children who haven't had chicken pox. The same holds for measles, mumps, and most childhood "infectious" diseases.[3]

Why do measles and chicken pox seem to be infectious? One child puts out the message through exosomes that now is time to go through the detoxifying experience called chicken pox. Other children in their home or class or town receive the message and begin the same detoxification experience. In the end, the children are all better off for having "sung" together.

With an illness like herpes, resonance may also be at play. (Also, a collagen deficiency may contribute to the genital irritations in patients with herpes and other STDs.)

So when two people come together in the highly charged act of sex, a situation in which this resonance acts strongly, it is no surprise that the couple might resonate together and create identical DNA or RNA, in a manner similar to what occurred in the beaker. To a virologist, this looks like the appearance of a new contagious virus. To a realistic observer, it is two people forging an intimate genetic connection. This observation, rather than proving contagion, teaches us about the mystery we call *life*. It teaches us again that the materialistic conception of the "wily attack virus" is an impoverished, inaccurate view of the world. And it teaches us to forgo simplistic explanations and look into the deepest mysteries of life if we are to create a world of health and freedom.

The discoveries about the resonant properties of genetic material can also help us explain how humans and animals adjust to new situations—a new toxin or new electromagnetic frequencies—not by competition and survival of the fittest but through the harmonizing of shared experience.

Imagine a situation where the human community is confronted with a new toxin. The new toxin can be neutralized only by an enzyme that is not usually made by human beings. But one member of the community has a randomly generated mutation that allows her—and only her—to make the toxin-neutralizing enzyme. She does well, whereas others sicken and some die because this randomly generated mutation gives her an adaptive advantage. According to the theory of genetic mutation and natural selection, her genes will slowly spread throughout the population. But what if she is a sixty-year-old postmenopausal woman, or a man who does not have children? Then the helpful gene would die out. If we're lucky, the carrier of the gene will be a thirty-year-old man about to get married. He and his wife have six children with three carrying the autosomal dominant mutation. One of those three dies in a car crash, the other becomes sterile following a Gardasil vaccine, and the third passes the adaptive gene on to her two children. In ten thousand years, that adaptive gene will have spread throughout the population through natural selection. Unfortunately, the toxin either has killed everyone off by then or is long gone, so the mutation is useless. It's clear that the theory of natural selection following random mutations cannot explain how humans and animals adapt to new situations in time for these mutations to be useful.

So how do we adapt? Our threatened cells produce exosomes containing DNA and RNA, which have a unique resonance. The pattern of this genetic material will quickly pass to others through resonance (especially if they are in close contact). This is the role of "viruses" in nature; they are physical-resonance forms of genetic material that code for changes happening in the environment. They provide real-time genetic adaption. It's a totally ingenious system that we have missed by assuming that viruses are hostile and dangerous. A war on viruses is nothing more than a war on the forward evolution of humanity.

PART 2

WHAT CAUSES DISEASE?

CHAPTER 8

WATER

If the practice of medicine were conceived properly in the Western world, doctors would begin by ascertaining four basic factors: the quality of the water their patients drink; the quality of the food they eat; the level and type of toxins, including mental and emotional toxins, to which they are exposed; and finally the level and type of electromagnetic fields to which they are subjected. The vast majority of medical problems can be understood by gathering patient information on these four areas, and the vast majority of health problems can be helped or even solved, by "remediating" these four core issues.

Water, especially the water that supports life inside our cells and tissues, has amazing properties. We are accustomed to think of water existing in only three states: solid, liquid, or gas. However, water—and only water—also has a fourth state, sometimes called coherent water, *structured water*, or simply the gel phase. Each phase of water has unique characteristics in terms of bond angles (the angles between the hydrogen and oxygen molecules), charge, motion characteristics, and many other physical properties.

Dr. Gerald Pollack, author of the groundbreaking book *Cells, Gels and the Engines of Life*,[1] along with biologist Dr. Gilbert Ling, were the first to describe the fourth stage of water and delineate its properties. Pollack coined the term EZ (exclusion zone) water. Fourth-phase water will structure itself against a hydrophilic ("water-loving") surface. Instead of moving randomly, the water molecules line up and form a crystalline structure that can be millions of molecules deep; this structure excludes every mineral and every other type of molecule or chemical from its midst. The water outside the EZ is "bulk" water, which contains minerals and dissolved compounds. It is basically "disordered," whereas EZ water is "ordered." Water is called the "universal solvent" because any hydrophilic substance will dissolve in it. EZ water is a pure crystalline "structure" consisting solely of hydrogen and oxygen.

EZ water is neither liquid nor solid, but rather is similar to a gel. To picture this fourth phase of water, imagine Jell-O (a coherent, non-solid, nonliquid, nongaseous mixture of unfolded proteins and water). The embedded water is arranged in clusters of molecules that organize themselves into a regular ("coherent") structure that we see as a gel. Jell-O is 99 percent water, yet water does not come out of Jell-O if it stays at the right temperature. Our bodies are 45–75 percent water, yet when we cut ourselves, water does not flow out because the water in our bodies is "structured" against the various hydrophilic surfaces in our tissues.

EZ water has a negative charge. In contrast, bulk water has a positive charge, making the water in our cells a kind of battery. The energy that charges the battery is heat and light energy ranging from the infrared through visible light through UV. This is why we feel better when we are in the sunlight, especially in the early morning or evening, which contains a lot of infrared light. This is why saunas (and heat in general) make us feel better. Heat and light help your intercellular and extracellular water form larger EZs. Fever does the same thing, which is why we should not suppress a fever.

Water from melting glaciers and from deep wells and springs are good sources of structured water because EZ water is created under pressure. Holy waters from the river Ganges and from Lourdes, with known healing properties, contain high amounts of structured EZ water.[2]

Recent studies have revealed that relaxed muscles contain mostly EZ water, whereas contracted muscles transition to mainly bulk water.[3] Anesthetics and drugs that reduce pain reduce the size of the EZ zones in our cells.

EZ water is the perfect "structure" for life processes because this fourth phase water gel can be shaped by the proteins, minerals, nucleic acids, lipids, and other substances in our body to form any shape or configuration of gel. This gel has an infinite number of binding sites, which allow it to change in response to a new stimulus. This stimulus can be in the form of chemicals such as hormones, energies such as thoughts and feelings, or even the resonant energies of the earth, sun, and stars. The shape of this gel unfolds the nucleic acids embedded in it, thereby controlling the expression of the genetic material. The structured EZ water in our cells, sometimes only a few molecules deep, is like a fine mesh of wires that carries energy and information.

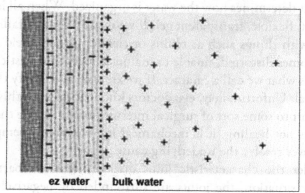

Against a hydrophilic surface (left), EZ water excludes all minerals and has a negative charge. Bulk water contains minerals and other compounds and has a positive charge. (Illustration from *The Fourth Phase of Water*, Ebner and Sons. Reprinted with permission.)

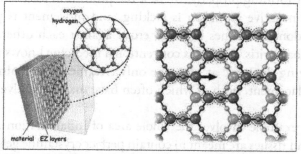

EZ water has a crystalline structure made up of layers of six-sided rings, slightly displaced. In ice, the layers of six-sided rings are not displaced. (Illustration from *The Fourth Phase of Water*, Ebner and Sons. Reprinted with permission.)

Toxins and EMFs damage the gel in our cells, interfering with virtually every physiological process. This damaging of the gels is a huge factor in disease; in essence, it is the unified field principle behind health and illness.

Consider the lens of the eye, one of the purest examples of a structured water gel in the body. The lens of the eye is a crystalline water structure, organized—as are all tissues—by a unique composition of proteins, lipids, minerals, nucleic acids, and other components. These components form the casing or housing for the crystalline water that forms the bulk of the lens. The lens must be transparent to light, and this requirement determines how the water is organized. When all is well, the lens is a soft, flexible, transparent gel. If we disturb the crystalline nature of the gel with things such as toxins or electromagnetic field exposure, the gel becomes distorted, unable to maintain its characteristic transparency. This is what we call a cataract. If we are able to detoxify the gel, the lens can heal. Unfortunately, eye doctors know nothing of this dynamic so they resort to some sort of surgical intervention to replace the diseased lens. This is not healing, it is mechanical intervention, a temporary fix that can never resolve the underlying cause.

Consider the characteristic joint disease called osteoarthritis. In the healthy situation, the joints are surrounded by negatively charged gels called bursa. These gels not only physically protect the underlying bones (themselves a denser type of gel), but, because they are negatively charged, when two opposing bursa come together the negative charges repel each other, thus ensuring smooth movement. When the gels are sick and not forming properly, we lack the protection for the underlying bones. The negative repulsion is lacking, and movement is painful. If nothing is done the bones begin to erode against each other, a process we call osteoarthritis. Again, as conventional medicine knows nothing of the underlying dynamic at play, the only treatments are pain medicine or joint replacement, both of which often have major negative effects for the patient.

A final example involves the whole area of inflammation and fevers. Our cells and tissues are meant to contain perfect crystalline gels. If a toxin becomes dissolved into the gels, the body attempts to rid itself of this toxin. The way the body does this is to elevate the temperature (we call this a fever), which partially liquifies the gels so that the toxins can be flushed out in mucus, after which we feel better, meaning we reconstitute our perfect gels once again. Fever and inflammation is simply a detoxification process, not a disease that needs to be suppressed.

Until doctors understand these simple principles, we must suffer under a medical system that cannot heal. That is one of the biggest tragedies of our time.

Jell-O is *fractal* in nature—meaning that any small piece of the gel has the same molecular shape as the larger gel. Examination of the smallest molecular units of the gel reveals it to be the same form as the macroscopic unit. This feature allows information to pass via all levels and to connect the molecular level to the macroscopic level. Here, we can only hint at the crucial importance of the coherent nature of water as the basis of life.

Preliminary findings indicate that when structured water is exposed to a Wi-Fi signal from a nearby router, the size of the EZ diminishes by about 15 percent.[4] This finding has profound implications for the interaction of EMFs and the structure of water in our cellular gels. If a nearby Wi-Fi router causes such a change, we can only imagine what the millimeter waves of 5G do to the structured water in our tissues.

Since human beings are made up of 70 percent water by volume and over 99.99 percent of the molecules in a human being are water molecules, we need to pay attention to the quality of the water we drink. Health professionals' foremost concern should be the type of water and other liquids their patients are consuming.

Water consumed by healthy nonindustrialized peoples had four characteristics: First, the water was free of toxins. This is in complete contrast with the municipal water that most people drink. Today's water contains chlorine and chloramine, which are toxic to our microbiome, as well as to the rest of our body. Today's water contains fluoride, an industrial waste that is toxic to the enzymes in our tissues, enzymes needed for every chemical transformation that happens in our bodies. Water also contains microplastics, which can line and congest the intestinal wall, and aluminum, which predisposes us to electrical sensitivity as well as a laundry list of diseases. Municipal water contains residues of many pharmaceutical drugs, including birth control pills, statins, and antidepressants.

Second, traditionally consumed water was abundant in vital minerals such as magnesium, calcium, zinc, and iodine.

Third, all traditionally consumed water was at least partially structured and organized into EZs because in nature, water moves in vortex patterns. Water bubbles up from the ground in springs, swirls in pools, flows over rocks, and forms eddies and vortices. Water flowing in vortex patterns becomes more "coherent," becoming increasingly structured. The structure actually persists for some time, and does not revert to incoherent bulk water just because it stops flowing.

Water flowing in vortices also picks up oxygen from the air and becomes more oxygenated. In contrast, most municipal water is stagnant in tanks and then flows through linear pipes with no vortex movement possible. This water is devoid of structure and coherence, and also completely devoid of dissolved oxygen; this lack of oxygen has a deleterious effect on our microbiome.

Finally, traditionally consumed water was exposed to the sounds and wavelengths of the natural world. Water flowing in mountain streams is exposed not just to the minerals, microbes, and other constituents of the forest, but also the sounds and energies of the life of the forest and of the entire natural universe, including the stars, sun, and moon.

It's clear that most water is sick and toxic and a major contributor to illness. If we are ever to change the course of the disease patterns in our world, it must start with clean, healthy water. Clean, healthy water should be completely free of any toxins: no chlorine, no fluoride, no aluminum, no lead, no pharmaceutical residues, no microplastics—nothing should be present that is not a natural component of water flowing in a healthy mountain stream.

Currently the only way this is possible is to purify the water down to the nanoparticle level. This is a monumental task but one that can be accomplished with the proper equipment. It is truly unfortunate, even tragic, that our water must be cleaned and "purified" in order for us to consume it, but until the world wakes up to the fact that poisoning the water is a completely unacceptable practice, we'll have to take these precautions.

Whole-house water purifiers do exist that can filter the water and then add minerals while oxygenating and structuring the water by letting it flow in a vortex pattern. There are also less expensive ways of creating clean, structured, well-oxygenated water (see Appendix A).

Drinking well-oxygenated water is important. In addition to their highly structured quality, the two most sacred and medicinal waters in the world, the water of the grotto at Lourdes (coming from a very deep spring) and the water in the Ganges river (coming from glaciers in the Himalayas), are likely highly oxygenated.[5]

High oxygen levels help explain why the waters of Lourdes and the Ganges have been associated with healing of a variety of diseases. Oxygen is essential and foundational for a healthy life; increasing the oxygen levels in our tissues improves the function and in particular the energy-generating capacity of our tissues. Oxygen deficiency has been widely associated with the development of cancer through the well-known Warburg effect; that is, the switch from aerobic to anaerobic fermentation processes in our cells. Hypoxia, the condition of low oxygen levels in the tissues, is a typical symptom of Covid-19.

Conventional researchers often claim that we can get oxygen into our bodies only though our lungs, so how do the oxygen levels in our water affect our health? Like so many other "truths" in science, the idea that we only absorb oxygen through our lungs is incorrect. If one uses sensitive oxygen measurement devices, one can demonstrate that soaking in a tub of highly oxygenated water and drinking highly oxygenated water will cause the oxygen level in the blood to rise.[6] This proves that at least some oxygen is absorbed both through the skin and the GI tract.

Highly oxygenated water contributes to our health in another important way, as shown in a study of plant growth. Research has shown that watering plants with highly oxygenated water stimulates their growth and improves their health and resistance to disease.[7] For many scientists, this makes no sense because we are told that plants don't use oxygen but rather they breathe out oxygen. How is it possible that exposing plants to oxygenated water increases their health and vitality?

The answer is clearly that the oxygen doesn't affect the plants directly but that the oxygen is used by the microbes in the soil. Watering plants with oxygenated water stimulates the growth of healthy aerobic bacteria in the soil. Plants don't primarily eat or absorb nutrients from the soil; rather, they (like us) eat the "waste products" of the bacteria in the soil. If we feed the microbes in the soil healthy nutrients including oxygen, the healthiest microbes will flourish. These put out the healthiest nutrients, which are absorbed by the plants to create healthy, flourishing plants.

So it is with us. We don't actually absorb the nutrients directly from our food, at least not solely. Rather, we eat food and drink water to nourish the billions of microbes in our gut. If we increase the oxygen level in the water we drink, we grow healthy aerobic bacteria in our GI tract. These aerobes use the water and the food we consume to make the highest-quality nutrients for us to absorb. And with plenty of oxygen, these healthy microbes will not switch to an anaerobic metabolism that produces toxins.

Life is a complex dance of nature, microbes, and organisms. Oxygenated water produces the conditions in which the healthiest microbes flourish and produce robust, vibrant, disease-resistant people, plants, and animals. Microorganisms that don't have enough oxygen become anaerobic and produce toxins that cause diseases like botulism, tetanus, cholera, and typhus.

Recent research indicates that drinking oxygenated water improves wound healing,[8] enhances lactic acid clearance in athletes,[9] improves immune status,[10] and protects against muscle fatigue.[11] Oxygenated water is a much better choice for athletes than steroids! In addition, conditions of low oxygen promote cancer growth.[12]

Currently most people consume devitalized food and oxygen-deficient water; antibiotic use is rampant and most people therefore have a predominance of toxic, disease-causing anaerobic bacteria in their GI tract. And, after all this we blame our illnesses on a virus that we can't even find!

The final step in the production of healthy water is the exposure, as in nature, of the water to the sounds and energies of nature. This can be accomplished through the exposure of the finished water to birds, frogs, trees, and other living things, or even through exposing water to sacred sounds, to music, to healing vibrations, to a blessing—or even to healthy, loving vibrations in the household. This crucial finishing step recreates the process through which nature "produces" water fit for consumption. All animals will naturally seek out such water. All animals when allowed free choice will eschew the toxic, dead, industrial water consumed by the vast majority of the people in the world.

As we will see in chapter 8, fermented beverages like kombucha and kefir achieve a kind of structure through the fermentation process and an effervescence that structures water around each bubble of air. Gelatinous bone broth carries the structure of collagen, which helps create healthy collagen throughout the body, down to the smallest structures in our cells—water structures itself against the hydrophilic surfaces of this collagen. The water we get in fruits and vegetables is also structured.

Well-oxygenated water for drinking and bathing should be the first thing patients receive when they enter a hospital or nursing home. Until then, see Appendix A for sources of healthy options, which are the best we can do as we go about the long and tedious process of reclaiming our world.

CHAPTER 9

FOOD

In the late 1890s, as the germ theory gained prominence, a new invention came on the scene: the *stainless-steel roller press*. This gleaming contraption allowed manufacturers to extract oil from hard seeds like corn kernels, cottonseeds, and soybeans. Primitive stone presses worked only for oily seeds like sesame, flax, and rapeseed and for oily fruits like olives, coconuts, and palm fruit. A traditional stone press extracts the oil slowly and without heat, so the final product is natural and tends to be healthy.

Oil from cottonseed—a waste product of the cotton industry—was the first fabrication of the new mechanized press. Like all industrial seed oils, cottonseed oil oozes out of the crushed seed as a dark, smelly gunk, something no one in his right mind would consume. High-temperature processing, which involves alkaline chemicals, deodorizing, bleaching, and hydrogenation (a process that turns a liquid oil into a solid) transforms the dark gunk into a product suitable for its initial use: candles. Proctor & Gamble, located in Cincinnati, perfected the refining process for this industrial product. But with electrification, the candle industry declined. What were they going to do with the expensive processing infrastructure in which they had invested? Feed the oil to people, of course.

The result was a profound change in the food supply, something the world had never seen. It took about forty years for industrial seed oils—as both hard, partially hydrogenated fats and as liquid cooking oils—to replace animal fats for cooking and baking; cheap industrial oils from cottonseed, corn, and soybeans made the processed-food industry possible—so cheap and so profitable that the industry had plenty of money for marketing campaigns and plenty of clout to influence university research and government policy. For years, health organizations including the World Health Organization (WHO) have recommended a diet containing industrially processed vegetable oils instead of natural animal fats.

No dietary change has ever been as deleterious to health as the advent of industrial seed oils, usually called "vegetable oils"; loaded with chemicals, intrinsically rancid, and lacking the many essential nutrients that humankind gets exclusively from animal fats like butter, lard, poultry fat, and tallow, they are a recipe for poor health. Chronic disease such as heart disease and cancer, kidney problems, Alzheimer's, and immune disorders have increased in lockstep with the increase in vegetable oil consumption. Moreover, the type of fat molecules in vegetable oil (omega-6 linoleic acid) can make our bodies more sensitive to the effects of electromagnetic radiation.

We have trillions of cells in our bodies, and every cell is surrounded by a membrane composed of a double layer of fat molecules, called the *lipid bilayer*. These molecules are mostly *saturated* because, after all, they are animal fats. The other main component of the cell membrane is *cholesterol*. Together, the saturated fats and cholesterol ensure that the cell membrane is waterproof, thus allowing a discrete chemistry and a different electrical potential inside and outside the cell. The remarkable membrane is engineered with channels and receptors so that only certain compounds get in and out.

Embedded in the cell's interior are the mitochondria, which help create energy. They are like tiny electric motors inside our cells. These too have a membrane composed of a double layer of fat molecules, most of which should be saturated, in order for the mitochondria to support efficient energy generation for our cells and bodies.

As we explained in chapter 8, the structures in your tissues serve to create microscopic enclosures where water structures itself against billions of hydrophilic surfaces. The areas of structured water have a negative charge. Inside the cell, structured water fills the spaces, creating what amounts to a web of fine wires to carry electric current through the cell and on to other cells. Good health depends on keeping this gelled structure protected and intact—protected from poisons, EMFs, and even negative emotions. The goal is to keep our own internal currents as safeguarded as possible against interference from 5G and other outside

EMFs. Saturated fats serve as a kind of insulation in cells and tissues. On the other hand, the types of fat molecules in vegetable oils—called polyunsaturated fatty acids—do not provide the stability these structures need. When built into our cell and tissue membranes, the cells become "floppy" and "leaky"; they can no longer provide effective barriers that our cells require to function properly.

Having adequate saturated fat in our cell membranes is especially important in the Internet age because 5G and other EMFs increase the permeability of the cell membrane,[1] which can result in a kind of starvation of all our tissues, with all sorts of unfortunate consequences—from fatigue to cancer.

At least half the fat molecules in the cell membrane need to be saturated in order for our cells to function optimally. The fat molecule in our lung surfactants needs to be 100 percent saturated for the lungs to work properly.[2] If our diet is lacking in saturated fat, the body will put polyunsaturated or partially hydrogenated fatty acids in the lung surfactants, making respiration difficult, with lung disease such as asthma and pneumonia likely consequences.[3] Chronic lower respiratory disease includes chronic obstructive pulmonary disease (COPD), emphysema, and bronchitis. The lungs simply cannot work properly in those who consume a lot of industrial seed oils.

Saturated animal fats also supply cholesterol, which is needed in the cell membranes to ensure the cells are waterproof so they can have a different electric potential on the inside and outside of the cell. Another important compound we get uniquely from animal fats is arachidonic acid, which is required for tight cell-to-cell junctures.

A key function of saturated animal fats is to serve as carriers for a trio of fat-soluble nutrients: vitamins A, D, and K2. The levels of these vitamins were much higher in the diets of our ancestors and of nonindustrialized peoples, for three reasons. First, most of the fats that our ancestors consumed were animal fats—butter, lard, poultry fat, and tallow. Second, they ate the whole animal—not just the muscle meats but also the organs, marrow, skin, and blood. The fat-soluble vitamins are concentrated in these organ meats, especially the liver. Even as late as World War II, Americans typically ate liver once a week, giving them a consistent dose of vitamin A.

Third, animals were raised on pasture in the sunlight, which serves to maximize the amounts of these key nutrients in our food. Egg yolk from a chicken raised outside, the old-fashioned way, contains several times more vitamin D than does an egg yolk from a chicken raised in confinement—the "modern" industrial model.[4]

Nothing can happen in the body without vitamins A, D, and K2—from growth to hormone formation, to energy production, to reproduction—this triumvirate of nutrients works together to protect us against toxins and enhance immunity. Vitamin A is particularly important for healthy lung function.[5] The best sources are cod liver oil, organ meats from healthy animals (think liver, liverwurst, scrapple, pâté, and terrines), egg yolks from pastured hens, butter and cream from grass-fed cows, fish eggs, shellfish, oily fish, lard from pigs raised outdoors, and poultry fat and poultry liver from birds raised in the sunlight on green grass—all food items that conventional public health officials discourage us from eating or that modern industrial agricultural practices make it difficult to obtain.

Modern eating practices rob us not only of these nutrients but of minerals as well, because the fat-soluble vitamins play a key role in mineral assimilation. Smoothies made from organic vegetables contain minerals, but these largely go to waste without the fat-soluble vitamins.

Industrial seed oil production fills our bellies but starves our cells; the same can be said for industrial grain production. A triumph of industrial processing is the Chorleywood method, whereby grains of wheat can be transformed into loaves of bread in their plastic bags in two hours; also the high-temperature, high-pressure extrusion process, which produces dry breakfast cereals like Cheerios and Wheaties out of wheat, oats, and corn.

Traditional, nonindustrialized cultures from around the world did not eat grains in this way; instead, they subjected them to a long, slow fermentation process, such as soaking oats overnight or even for several nights before cooking them into a sour porridge.[6] Naturally leavened sourdough bread is a fermentation process that takes several days. In Africa, parts of the Middle East, and also in medieval Europe, slow fermentation of grains was the first step in creating nourishing beverages like sorghum beer and small beer—beverages of low alcohol content and high levels of nutrients, especially B vitamins. Small beer was a common beverage, even for children, in colonial times—Benjamin Franklin consumed it for breakfast, and George Washington had a recipe for small beer involving bran and molasses.[7] Such a beverage would have nourished the gut flora by providing structured water around the fizzy bubbles and B vitamins galore.

Grains that haven't been soaked, sprouted, or fermented are difficult for humans to digest and contain many "anti-nutrients," compounds like phytic acid, lectins, and enzyme inhibitors, which block digestion and can even lead to mineral deficiencies. Modern grain products—including trendy "health" products like oat bran muffins and granola—fill the belly but do not nourish. Sometimes they even poison. The extrusion process used to make breakfast cereals creates neurotoxins;[8] gluten in wheat becomes toxic without proper preparation.

Careful preparation transforms grains into real food—increasing B vitamins and liberating minerals for easy assimilation. The food industry "solves" the problem of modern grain processing by adding synthetic vitamins. In any event, overt deficiencies are rare in America, not because of the synthetic vitamins added to grains, but because most Americans eat plenty of meat.

An interesting symptom in some Covid-19 patients is "Covid toes"—red, inflamed toes, similar to pellagra toes (which is caused by a niacin deficiency).

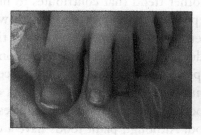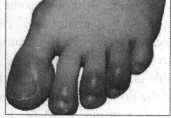

Pellegra Toes (L); Covid Toes (R)

Scientists have observed a threefold depletion of NAD (a form of niacin) in the cells of Covid patients, a condition blamed on the coronavirus. However, exposure to wireless technology and microwave radiation can also deplete cellular forms of niacin.[9] The obvious defense is to limit EMF exposure and consume ample B vitamins, especially niacin. Proper preparation of grains and sufficient animal products will ensure adequate levels of the B vitamins.

Other obvious products of the Industrial Revolution are refined white flour and sugar (and its modern evil twin, high-fructose corn syrup [HFCS]). Refined sweeteners and white flour are the quintessential "displacing foods of modern commerce." Ironically, these refined foods actually play an important role in the diets of those avoiding animal fats. The body has such a great need for saturated fat to maintain the membranes and surfaces in the tissues that it has a backup plan in case our diets do not contain adequate amounts: it makes saturated fat from carbohydrates, especially refined carbohydrates.[10] Unfortunately, this backup plan will not provide the fat-soluble nutrients we get from animal fats nor the B vitamins in their natural forms that we get from properly prepared whole grains. Instead, consumption of refined carbs serves as a fast track to the chronic diseases from which Westerners suffer—diabetes, heart disease, kidney problems, high blood pressure, and cancer. The vast majority of Covid-19 victims suffer from one or more of these preexisting conditions.

Diets based on vegetable oils or even on olive oil often lead to cravings for refined carbohydrates; a return to animal fats is the first step to resolving the need for refined carbs.

The embrace of "plant based" diets (whether vegan, vegetarian, or just low in animal products) is another trend that contributes to nutritional starvation. Although some people report improved health when they embark on a "plant based" diet, deficiencies will develop over time. The decision to avoid animal products often comes with resolutions to "eat better" in general and avoid processed food. Removing sources of vegetable oils, white flour, and refined sweeteners from the diet is only part of the process for regaining good health; the other part requires the consumption of nutrient-dense food. Although plant foods have a definite role to play in human diets, they are far less dense in vitamins and minerals than are animal foods.

Long-term vegetarianism, especially veganism, often results in deficiencies of complete protein; the fat-soluble vitamins A, D, and K2; vitamin B12; and four key minerals: zinc, sulfur, iron, and calcium. On the other hand, plant foods like beans, nuts, and grains tend to be high in copper, and a high copper-to-zinc ratio can dispose one to electromagnetic sensitivity.[11] Lack of iron, of course, leads to anemia and fatigue; sulfur supports mechanisms for oxygen transport in the blood. Both zinc and sulfur supplementation seem to help Covid-19 patients. Best sources are animal foods like red meat, liver, and egg yolks. Vitamin A from animal fats and liver helps ensure that iron goes into the red blood cells where it is needed and that all minerals are used effectively. The body has a harder time using iron added to processed foods, like breakfast cereals and white flour, so it ends up in the soft tissues where it does not belong—so-called toxic iron—rather than in the bloodstream where the iron in red blood cells carries oxygen to the tissues.

Covid-19 and zinc deficiency have many symptoms in common: cough, nausea, fever, pain, abdominal cramping, diarrhea, loss of taste and smell, loss of appetite, fatigue and apathy, inflammation, and decreased immunity. Zinc-rich foods and even zinc lozenges provide real protection against this disease.

One effect of 5G seems to be the stimulation of calcium channels in the cell membrane. This drives calcium into cells, essentially poisoning the cell, while lowering the ionizable calcium in the blood. The ionized calcium in the blood is used in the coagulation pathways to help clotting and prevent uncontrolled bleeding. If it drops too low, people hemorrhage. During the 1918 pandemic, many doctors noted that their patients died from hemorrhage, not pneumonia. Some doctors reported that IV calcium lactate kept people from dying. Soon after, Royal Lee from the Standard Process company formulated a flu product called Congaplex, which contained calcium lactate—the same form of readily available calcium as in raw milk. In addition, raw whole milk from pastured cows contains compounds that strengthen the immune system and help us deal with stress and EMFs.[12]

The greatest tragedy of the germ theory has been its application to milk, nature's perfect food. Today, most milk is subject to pasteurization; in fact, most milk is ultra-pasteurized, a process that flash heats milk to 230 degrees Fahrenheit, way above boiling point, ostensibly to rid milk of harmful germs, but actually to lengthen shelf life. Unfortunately, pasteurization greatly lowers the vitamin content—one dairy industry study found that pasteurization resulted in lower amounts of all the B vitamins, especially B2, B6, and B12—and these were studies of milk that was merely pasteurized, that is, heated to 170 degrees.[13] It's likely that ultra-pasteurization results in the destruction of up to 100 percent of the vitamins in milk. The minerals will remain, but the enzymes the body needs to assimilate these minerals are destroyed. Pasteurization destroys beta-lactoglobulin, which is needed for the intestinal uptake of vitamins A and D.[14]

Pasteurization in the name of the germ theory has resulted in the destruction of most of the goodness in milk, a prime food for growing children in Western culture. Pasteurization also renders milk proteins allergenic; many people with milk allergies turn to "milk" made of almonds, peas, oats, or soy, which have dubious nutritional value.

Raw, whole milk (especially from grass-fed animals) is a complete, easily digested food. It contains every nutrient that babies and children need for growth; it protects them from asthma and respiratory illness;[15] it ensures plentiful calcium that is easily assimilated for strong teeth and bones. For the elderly, raw milk is equally nourishing; it protects the bones and nourishes the tissues, even when the digestive fires have waned.

Raw milk is an excellent source of glutathione, a compound our bodies use for detoxification. Only glutathione from fresh, denatured whey proteins will work—that means from raw milk, not from pasteurized milk or whey powders. Alexey V. Polonikov of Kursk State Medical University proposes that "glutathione deficiency is exactly the most plausible explanation for serious manifestation and death in Covid-19 infected patients."[16] Raw milk can be of immense help in protecting us from this disease.

Another important source of nutrients missing from modern diets: gelatin-rich bone broth made from the bones and cartilaginous portions of the animal, which nourishes the cartilage in our own bodies—and our bodies contain more cartilage than muscle. Bone broth is rich in glycine, an essential element in collagen that helps maintain structured water inside and outside of cells. Glycine helps create strong collagen in certain types of lung surfactants and throughout the body, and it supports detoxification.

Animal feet, heads, bones, and skin did not go to waste in your grandmother's kitchen. They were thrown into a pot and simmered on the back of the stove to make a rich broth—basically melted collagen. This broth then formed the basis for nourishing soups, stews, sauces, and gravies—or given as just a mug of broth for optimal energy and good digestion—a much better choice than coffee! Unfortunately, the food industry has figured out a way to imitate homemade broth—the broth-based gravy that your grandmother served appears in imitation form as a kind of goop made with water, a thickener, artificial colorings, and artificial flavorings, especially monosodium glutamate (MSG), a neurotoxin. MSG appears in many canned and dehydrated soups and stews, bottled sauces, "broth" in aseptic containers, salad dressings, seasoning mixes, soy foods (which are intrinsically bitter), and even vegetable oils. Rarely labeled, MSG is a neurotoxin, not a nutrient, and another source of starvation for those who consume mostly processed food.

Another important component of traditional nutrient-dense diets is fermented food and beverages. Raw fermented foods supply beneficial bacteria to the intestinal tract, preferably on a daily basis. These bacteria help digestion, liberate minerals, break down anti-nutrients, supply vitamins (especially B vitamins), and protect us against toxins. In fact, a recent study links fermented vegetable consumption to low Covid-19 mortality.[17] Fermented condiments like raw pickles and sauerkraut, fermented sauces like ketchup, and fermented beverages like kefir and kombucha are critical components in a diet that truly nourishes and protects. Unfortunately, the modern diet replaces raw fermented condiments with canned versions, makes heat-treated ketchup loaded with additives, and promotes truly toxic and heavily sweetened soft drinks instead of artisanal fermented beverages.

The public first learned about the benefits of lactic acid–producing bacteria in fermented foods, especially fermented milk products like yogurt, from bacteriologist and Nobel Prize winner Ilya Mechnikov, a contemporary of Louis Pasteur. Mechnikov is credited with the discovery of macrophages, which turned out to be the major defense mechanism in our innate immune system. He proposed the theory that white blood cells could engulf and destroy toxins and bacteria, which met with skepticism from Pasteur and others. At the time, most bacteriologists—always assuming that natural processes are detrimental—believed that white blood cells ingested pathogens and then spread them further through the body.

Unlike Pasteur, who believed that all bacteria were bad, Mechnikov credited the good health and longevity of Bulgarian peasants to their daily consumption of (fermented) yogurt and the lactic acid–producing bacteria it contained.

Mechnikov, a colorful and passionate figure, twice attempted suicide—the first time by an opium overdose and the second time by injecting himself with the spirochete of relapsing fever (akin to malaria).[18] He concluded that it was his habit of eating Bulgarian yogurt that protected him against the spirochete toxins and allowed him to survive. He also experimented on himself and others by drinking cholera bacteria during the 1892 cholera epidemic in France. He and one volunteer did not get sick, but another volunteer almost died. He then discovered that some microbes hindered the cholera growth, whereas others stimulated the production of cholera toxins. He concluded that the proper cultivation of intestinal flora could protect against deadly diseases like cholera.[19] We get these protective bacteria on a daily basis when we eat lacto-fermented food.

An important component of fermented foods is vitamin C. Successful treatments for Covid-19 cases include large doses of vitamin C (oral or IV). Your best dietary source is fermented veggies like sauerkraut, which is manyfold richer in vitamin C than is fresh cabbage.

The food-processing technology that accompanied the Industrial Revolution has allowed us to ruin just about every common ingredient that we put into our mouths, even salt. Salt is a critical nutrient for health, and especially for maintaining the difference in electrical potential in our tissues, which can protect us against EMFs. But modern processing removes all the magnesium and trace minerals from salt, and it adds an aluminum compound that prevents clumping—so your salt will pour when it rains. The solution is to use unrefined salt on your food and in your cooking, salt that contains a wealth of trace minerals and provides us with a daily source of magnesium. One and one-half teaspoons of unrefined salt (the minimum adult requirement for sodium and chloride) actually provides about twice the minimum adult requirement for magnesium.

Now take your diet of processed food—your frozen dinner, your canned soup, your carryout meal, and your leftovers—and zap it with the microwave. Little that nourishes will be left in such food-like substances.[20]

The impression created in numerous books and in the media is that a "healthy" diet is dry and unsatisfying—containing skinless chicken breasts, lean meat, vegetable juices, and rough whole grains—could not be further from the truth. A healthy diet requires no sacrifice in taste and satisfaction, only care in purchasing and preparing our food. Rich, whole raw dairy products including plenty of butter, fatty meats, natural bacon and charcuterie, eggs (especially the yolks), artisan sourdough bread, genuine bone broth, satisfying sauces, natural sweeteners, plentiful unrefined salt, interesting condiments, and refreshing fermented beverages—these provide the kind of diet that truly nourishes and protects—and they are becoming more and more available commercially. If your diet contains mostly processed foods, you can be assured that your body is in starvation mode—especially if you heat up your food in a microwave oven.

For practical suggestions on adopting a nutrient-dense diet, see Appendix C.

CHAPTER 10

TOXINS

In chapter 3, we looked at some of the toxins that cause disease—mistakenly attributed to microbes—in times past. Those dwelling in cities and towns lived in constant danger of exposure to poisonous gases from sewage and manure. Volatile compounds like hydrogen sulfide, ammonia, methane, esters, carbon monoxide, sulfur dioxide, and nitrogen oxides can kill people through asphyxiation when they are exposed to high concentrations. Other effects include eye irritation, nausea, and breathing difficulty.

Our ancestors were also exposed on a daily basis to dioxins and other toxins in smoke—from fires built for warmth, cooking, and metal work. Even today, smoke from cooking fires is a major source of air pollution in the developing world, especially smoke from open fires inside houses and huts.

Our ancestors also had to contend with toxic metals: lead used in pipes, cooking vessels, building materials, and cosmetics; arsenic used in metal alloys, cosmetics, and medical treatments; and mercury used in ointments and medicines, cosmetics, metal amalgams, and silver mining.

Although today we recognize the extreme toxicity of these substances, they have not gone away, especially arsenic. Contamination of groundwater with arsenic is a problem that affects millions of people across the world.[1] Heavily used as an insecticide in the late nineteenth and early twentieth centuries, arsenic use in produce production is no longer common (except in cotton farming); however, its use as a feed additive in poultry and swine production, especially in the United States, has increased. For example, an arsenic-containing feed additive called roxarsone is used by about 70 percent of US poultry growers.[2]

The first edition of *The Merck Manual* featured many mercury-based medicines. A "tonic" containing mercury constituted the standard treatment for syphilis, the classic example of a deadly medicine far worse than the disease it was designed to treat. Use of mercury in medicine has declined, although some over-the-counter drugs including topical antiseptics, stimulant laxatives, diaper-rash ointment, eye drops, and nasal sprays contain mercury compounds. The main medical uses for mercury today are in dental amalgams (which constantly outgas mercury into the mouth and sinus cavities[3]) and as a vaccine preservative called thimerosal.[4]

After studies showing that the amount of mercury in the childhood vaccination schedule recommended by the CDC exceeded all national and global maximum safety limits, public pressure forced manufacturers to remove thimerosal or reduce it to trace amounts in all US vaccines recommended for children six years of age and under. But manufacturers did not remove mercury from the multi-dose vials of the inactivated flu vaccine, which is given to pregnant women, children under the age of one, and, of course, adults. The result is that if the flu vaccine is added to a child's vaccine schedule, he is likely to get more mercury in his bloodstream than he did before the removal of thimerosal from the other childhood vaccinations.[5] Small amounts of mercury residues from the manufacturing process still remain in most vaccines.

The industrial use of mercury has declined, but it is still an ingredient in some measuring instruments, and it fills fluorescent lamps, including those compact bulbs that have largely replaced incandescent light bulbs. If they break in your house, you and your family will be exposed to toxic mercury fumes; if they break in a landfill, the mercury will pollute the soil and groundwater.

Cyanide compounds are a by-product of many industrial processes, such as oil refining and the production of polyurethanes. Many cyanide compounds are toxic; they can prevent the production of ATP, needed for energy-making processes, particularly affecting the central nervous system and the heart, leading to hypoxia (oxygen deprivation), a common symptom of the Covid-19 illness.[6] Cigarette smoke is an undisclosed source of cyanide compounds.[7]

Humanity is also exposed to formaldehyde, benzene, cadmium, phthalates, fluoride, and chloride compounds in drinking water (including chloramines, preferred by public water agencies because they persist and do not break down over time), and a host of pesticides, including the highly toxic cholinesterase inhibitors (nervous system poisons) sprayed on citrus fruit. These make their way into milk, butter, yogurt, and cheese via citrus peel cake fed to dairy cows.

Toxins in food range from propylene glycol (an antifreeze)—added to ice cream to keep it soft and creamy; formaldehyde and methyl alcohol (breakdown products of the artificial sweetener aspartame); the artificial sweetener acesulfame-K; dough conditioners; artificial colorings and flavorings (including MSG); preservatives; artificial vitamins (including beta-carotene); and chemical antioxidants like butylated hydroxyanisole (BHA), propyl gallate, and tert-butylhydroquinone (TBHQ), which are added to vegetable oils and fried foods such as potato chips.

Those with preexisting conditions like diabetes, obesity, hypertension, and heart disease—those who are most vulnerable to the illness attributed to coronavirus—are likely to have developed these conditions in large part due to processed foods loaded with these additives. How much they contribute to this disease is impossible to say, but it is reasonable to speculate that processed foods containing these and other chemicals play a role as cofactors in the coronavirus illness, indeed in any disease. The standard American diet (SAD) not only starves our tissues, but it also poisons them at the same time.

Researcher Stephanie Seneff, PhD, has pointed out the fact that the early epicenters of Covid-19 correspond with areas of high air pollution—Wuhan, China; the tristate region (New York, New Jersey, Connecticut); northern Italy; Spain; and Jefferson Parish, Louisiana—and particularly with the use of biodiesel. She notes that a Harvard Institute for Public Health study found a strong correlation between exposure to particulate air pollution and Covid-19 deaths.[8] The researchers found that an increase of only one microgram per cubic meter in fine particulate matter was associated with a 15 percent increase in the Covid-19 fatality rate. It bears repeating that atmospheric dust (that is, pollution) can exacerbate the effects of EMFs.[9]

Seneff notes that New York relies heavily on biodiesel fuel for public vehicles, and New York State has a large number of manufacturing plants where biomass from various sources, including used cooking oil, is processed into biofuels. The state also encourages the use of biofuel for home heating oil.

Biodiesel and biofuel made from plants contain the herbicide glyphosate (Roundup), which Seneff believes has a unique mechanism of toxicity. She points to a case of a mechanic who tried to clean a clogged applicator for glyphosate-based herbicide using a bucket of diesel fuel as a solvent. He quickly developed a bad cough and started coughing up blood. Rushed to the hospital, he was diagnosed with pneumonitis, an inflammatory disease of the lungs caused by exposure to toxic substances.[10]

Seneff postulates that the organic molecules in diesel fuel enhance the uptake of glyphosate in the lung cells by acting as a surfactant.[11] Glyphosate substitutes for the amino acid glycine found in cartilage, numerous enzymes, and important lung surfactants, leading to a myriad of problems, including lung disease.

Seneff points out that many US Covid-19 hot spots include major ports such as Seattle, Los Angeles, New Orleans, Boston, and New York. She notes that air pollution from ships is more toxic than air pollution from land vehicles, because ships use the lowest grade of diesel fuel.[12]

In Europe, over 20 percent of automobiles are powered by diesel fuel versus 2 percent in the United States. Unable to keep up with the demand for diesel fuel, Europe imports biodiesel (made primarily from GMO Roundup Ready soy) from Argentina. The Chinese produce biodiesel from canola (rapeseed), which is highly sprayed with Roundup, much of which grows along the Yangtze River, which cuts through Wuhan.

In the United States, three cities have adopted biodiesel for vehicle use on the roads—New York (which powered eleven thousand vehicles at least partially on biodiesel as of 2017), New Orleans (which uses biofuel in buses), and Washington, DC—all Covid-19 hot spots. All crops used for biodiesel in the United States are Roundup Ready crops sprayed with glyphosate—corn, soy, canola, and hardwood trees.

Aviation biofuel is another potential source of airborne glyphosate, first introduced by United Airlines and now in use by at least four airlines using New York City airports. The New York borough most affected by coronavirus is Queens, on the flight paths of New York's three major airports (La Guardia, JFK, and Newark) and intersected by three major interstate highways (I-278, I-495, and I-678).

In Britain:

> News reports have singled out bus drivers and people living in the town of
> Slough (adjacent to Heathrow Airport) as being especially affected. Test
> flights and commercial flights running on aviation biofuel blends have
> been flying into and out of Heathrow since 2008. On the ground, the

mayor of London reported in July 2017 that about a third of the city's nearly ten thousand buses were running on 20-percent-blends of biodiesel; the mayor also stated that by 2018, London would no longer add pure diesel double-deck buses to its fleet.[13]

Taiwan has a low rate of coronavirus deaths. Taiwan's cities have plenty of air pollution, but not from biofuels; Taiwan vehicles do not use biodiesel. In May 2014, the state-run oil refining company started phasing out biodiesel fuel production because due to the island's high humidity, even a 2 percent biodiesel blend resulted in the growth of microbes that clogged fuel tanks.[14]

One New England hot spot is Chelsea, Massachusetts, where much biofuel is produced. As of May 1, 2020, Chelsea ranked number one in Massachusetts, with 363 cases per ten thousand people. Brockton, the number-two city, had only 185 cases per ten thousand.[15]

Glyphosate exposure comes not just through the air but in our food, and exposure is highest in the United States, which uses the most glyphosate per capita of any country. Seneff attributes the high rate of many chronic diseases, including diabetes, obesity, fatty liver disease, heart disease, celiac disease, inflammatory bowel disease, hypertension, autism, and dementia to glyphosate exposure. In a 2014 landmark study, Swanson and coauthors showed that many of these chronic diseases are rising in the US population exactly in lockstep with the rise in glyphosate usage, particularly on wheat, which is sprayed with glyphosate shortly before harvest as a desiccant.[16] Whether ingested in food or breathed in from biodiesel, the effects of glyphosate are insidious, cumulative, and widespread.

According to Seneff, glyphosate's

mechanism of toxicity has to do with a proposed ability to mistakenly substitute for the coding amino acid glycine during protein synthesis. This is plausible because glyphosate is a glycine molecule—except that there is an additional attachment (a methyl-phosphonyl group) to the nitrogen atom of glycine, which changes the size and chemical and physical properties of the molecule but does not prevent it from incorporating into a peptide chain. It can be predicted that certain proteins will be affected in a devastating way if glyphosate should substitute for particular glycine residues known to be very important for their proper function. I have found that many of the diseases with rising prevalence can be explained through glyphosate substitution in specific proteins known to be defective in association with those diseases.[17]

Starved and poisoned, the typical American soon develops one or more chronic diseases and seeks medical advice; once in the clutches of the medical establishment, he becomes the target of more toxins, starting with statin drugs. The list of side effects from statin drugs is long and includes muscle pain or cramps, fatigue, fever, memory loss, confusion, diabetes, kidney and liver damage, heart failure, and digestive disorders. Most seriously, statins diminish the cholesterol available to cells and diminish the fat-soluble vitamins and other nutrients that are carried in the lipoproteins. With statins, your cells are starved of the nutrients they need to produce energy and keep your intercellular water organized. A Wenzhou Medical University study found that Covid patients had significantly lower cholesterol levels than did controls.[18]

In addition to statins, most Americans take other drugs. A study published by the Mayo Clinic[19] found that 70 percent of Americans take at least one prescription medication and that 20 percent of Americans are on five or more medications. These include metformin to lower blood sugar, blood pressure medications including ACE inhibitors (which act on the same receptors as exosomes), steroids, antiepileptics, antidepressants, pain medications, stomach acid inhibitors, and antibiotics. Many also take one or more over-the-counter drugs such as Tylenol, cough suppressants, sleep aids, and antacids. All these drugs have side effects, which means that all of them can act as poisons in the body. The side effects of ACE-inhibitor blood-pressure-lowering drugs, such as Lisinopril, are similar to those of Covid-19: a dry, persistent cough, dizziness, possible nausea, headache and trouble breathing.[20]

A final toxin, one that highly impacts our response to electromagnetic radiation, is aluminum, with an electrical conductivity only slightly less than that of copper. In fact, there are few more biologically reactive metals than aluminum. Aluminum binds strongly to oxygen-based compounds such as the phosphate groups in ATP—needed for the production of energy. In simple terms, too much aluminum in the body reduces our energy.

Human exposure in the twenty-first century is especially high. Aluminum is in most public tap water—it is used as a flocculant to clarify the water, and it is discharged from fertilizer and aluminum manufacturing. Jet engines spew aluminum ions into the air, particularly problematic for those living in the flight paths of major airports.[21] Aluminum compounds abound in toothpaste, mouthwash, soaps, skin care products, tanning creams, cosmetics, shampoos, hair products, deodorants, baby products, nail polish, perfume, food, food packaging, sunscreen, antacids, and buffered aspirin. Aluminum levels are especially high in infant formula, particularly soy formula.[22] Aluminum leaches into food from aluminum foil and cookware.

Another undisclosed source is marijuana. Users can absorb as much as 3,700 micrograms of aluminum per joint, representing "a significant risk factor for neurodegeneration."[23]

Aluminum levels are especially high in the brains of those with Alzheimer's and autism.[24]

The body has a certain tolerance for aluminum—beneficial gut flora can prevent its absorption and a good immune system offers some protection against airborne aluminum. But the body lacks such tolerance for aluminum injected into the bloodstream.

Mercury may have been removed or reduced in vaccines, but not aluminum. In fact, manufacturers have added more aluminum in order to provoke the production of antibodies, said to prove an "immune response." All of the diphtheria, tetanus, and pertussis vaccines (DT, DTaP, Td, Tdap, and combination vaccines with a DTaP component), *Haemophilus influenzae* type b (Hib) vaccine, hepatitis A and B and the hepatitis A/B combination vaccines, the meningococcal and pneumococcal vaccines, and the human papillomavirus (HPV) vaccines contain aluminum.[25] In fact, the newest HPV vaccine (Gardasil-9), recommended for teenage girls and boys, contains more than double the amount of aluminum than the original Gardasil vaccine.[26]

In 2011, distinguished immunologist Yehuda Schoenfeld and colleagues proposed the term "autoimmune/inflammatory syndrome induced by adjuvants" (ASIA) to describe the unusual immune-mediated diseases in humans and animals that appear after injection with aluminum-containing vaccines. ASIA manifests as "vague and sundry symptoms—chronic fatigue, muscle and joint pain, sleep disturbances, cognitive impairment, skin rashes and more."[27] Aluminum "accumulates, and the more you put in the system, the more you have. When you inject aluminum, you inject it directly into the immune system."[28] The researchers also noted that a person would have to eat "one million-fold higher aluminum to get the same level of [injected] aluminum adjuvant at the level of the immune cells."

The amount of aluminum injected into babies via multiple vaccinations exceeds anything that can be considered safe. A baby who receives the recommended eight doses of vaccine at the two-month checkup receives 1,225 mcg of aluminum at once; fully vaccinated babies receive 4,925 mcg by eighteen months. The maximum allowable aluminum (considered safe) per day for intravenous parenteral feeding is 25 mcg.[29]

Flu shots given to seniors do not contain aluminum, but the multi-vial shots contain mercury; and all flu shots contain contaminants such as formaldehyde and polysorbate 80. Those who received the flu shot in the United States during the 2017–2018 influenza season had a 36 percent increased risk of coronavirus illness.[30] In northern Italy, a campaign to inject the elderly with new types of influenza vaccines took place in 2018–2019,[31] and in June 2019, the Chinese instituted mandatory flu vaccine for all ages.[32]

We live in a toxic world. Add millimeter wave 5G technology to the mix and illness is sure to ensue.

CHAPTER 11

MIND, BODY, AND THE ROLE OF FEAR

Scientists have made two grave mistakes in their centuries-long quest to understand the human mind. Once we see through these mistakes, a more realistic concept of the "mind" can emerge.

It's important to understand the mind—what it is and how it works—because the mind has a large role to play in the experience of "contagion." In other words, if we don't explore the nature of the mind and come to a realistic understanding of how the mind operates, we will fail to understand the concept of contagion in general and the experience of Covid-19 in particular. This is because fear, hatred, and lies are key components of the phenomena we call "sickness"; these negative emotions and behaviors seem to be "contagious," and they are present in the world at almost unprecedented levels at this time. It's time to integrate the concept of the mind into a realistic framework for health and disease.

The first mistake concerning the mind that scientists and researchers have made in past centuries is the assumption that physical matter is the only thing that exists. If this is the underlying assumption, then it is only natural to look for a "physical" site in which this mind resides and then attempt to understand how the anatomy, chemistry, or physiology of this organ's cells create the mind. Scientists have localized the seat of the mind in the brain. They postulate that the brain is made of physical matter—chemicals and atoms—and therefore these brain cells must somehow "secrete" the mind. The mind must be a physical, biochemical product of the brain, much as thyroid hormone is a physical secretion of the thyroid gland. But try as they will, they just can't come up with the chemical or group of chemicals that constitutes this mind secreted by the brain. As always, we are told it is just a matter of time, and of course, more funding, for scientists to solve this puzzle.

This "matter" called the brain is either 99.99 percent empty space (if it exists as a particle) or just wave energy (if it exists as a wave). To complicate matters, these same scientists tell us that what determines whether the stuff that makes up our brains is in the particle or wave form is how the "mind" of the scientist is observing the stuff. In other words, this mind, which can't be found, actually determines the form of the organ that is supposedly creating the mind. Thus, the scientists are caught in a Gordian knot. Like a rat trapped in an infinite maze, there is no escape from this central dilemma. The result is that scientists try to understand more and more of the conundrum's details, while never getting to the core of the issue. This is the central paradox of materialistic science and its offshoot, materialistic medicine.

Most neuroscientists attempt to find the source of the mind within the organ they postulate is creating the mind—our brain. This is akin to attempting to localize the source of the sound emanating from a radio by dissecting the radio into its component parts. Although a radio is needed to receive and play sounds, no one could possibly think the sound originates in the radio. The radio is a receiver, and the more in tune it is to the various waves and frequencies in the world, the better it can function as a receiver. A perfect radio would theoretically be able to pick up any radio wave signals anywhere, if it were powerful enough and tuned enough. Different sizes and types of radios have different abilities to pick up the various signals; no one claims that because one radio is small and old and picks up only local, strong signals that the other signals don't exist. It's clearly all about the power and clarity of the radio one is using.

So it is with the connection between the brain (as the presumptive "site") and mind. The brain is a receiver; it works in conjunction with the entire organism in a complex dance we call life. The body brings nourishment to the brain; it removes waste products from the brain; it connects the brain to the senses and to the fingers so they can touch objects and provide the brain with the information it needs in order to work.

There is no body–mind duality; that is a superstition of materialistic scientists. There is a human being, divided up into various water compartments, each working together to create this experience we call life.

The input to this experience, as in a radio, comes from the world—actually the universe—as a whole. Our bodies (with their respective minds) are the receivers for this input in the form of electromagnetic waves. And, since we now know that every "substance" is also its own form of wave, there is a natural resonance created when the energy of the world meets the organized water crystal known as the human being. The result of this resonance is output, otherwise known as thoughts, feelings, and actions. The mind is simply a made-up concept for this dance of life—input from the world, received in the form of resonance by our organism, leading to the creation of output in the form of thoughts, feelings, and actions.

The second mistake is failing to understand the role that water, in its crystalline nature, plays in this resonance phenomena. One important clue that water is the crucial element in the creation of this mind we call the human organism is that the organ that serves as the primary receiver of thought waves—the brain—is also the organ with the highest water content—it is 80 percent water by volume (about 10 percent higher than other organs). Not only is the brain the organ that has the highest amount of crystalline water, it even floats in organized, structured cerebrospinal fluid (CSF) thereby achieving the remarkable feat of overcoming gravity. Without this salty CSF bath, the brain would feel too heavy to carry around; the brain would push against the skull, thereby cutting off its own circulation.

Life is a manifestation of the force of levity—plants grow up, sap rises in trees, and animals stand and walk; in contrast, mineral substances succumb to gravity.

Life would be impossible if "gravity ruled" in the human being. Fortunately, the brain floats upward in its pool of salty, crystalline fluid. Levity, like a plant growing up toward the sun, is the basic expression of living beings. This organized water known as the brain, floating in its bath of crystalline fluid, serves as a perfectly tuned receiver for the thoughts of the world. This is an exact description given by the most sophisticated scientists, inventors, musicians, artists, writers, and poets. Universally, they describe the experience of working on themselves in one way or another—thinking, practicing, studying—and then one day, the inspiration or thought simply comes to them to become Newton's theories, Beethoven's ninth symphony, or Isaac Singer's sewing machine. We have all had the experience of receiving a thought and knowing it is right, or having pondered a question for many years and suddenly receiving the answer. Somewhere, somehow, thoughts (all these various wave forms) exist. It's just a matter of tuning our watery brains to pick them up clearly.

This brings us to "Covid-19" and the experience of fear. Whether by design or by accident, humanity is currently bathed in the waveforms of fear, hatred, and lies. No sensible person could possibly deny this. People don't know whom to believe, which news report is accurate, which scientists or government officials are lying or which are telling the truth. We have been told to fear and suspect each other as carriers of deadly germs; any and all differences between us, even things as superficial as the color of a person's skin, are grounds for even more fear, suspicion, and hatred. It is not hyperbole to claim that every single person on earth is now bathed in this sea of fear, hatred, and lies. This is what is picked up by all living organisms as the predominant waveforms in the world.

Naturally, these waveforms create physiological responses in our bodies, such as activation of our inflammatory systems, as we attempt to rid ourselves of these destructive thought patterns. Our cortisol production increases, adrenaline soars, blood flow constricts, and pupils dilate as we prepare to escape this danger. We have been poisoned, drowned in this toxic brew, deep into our watery structures.

We also know that exposing organisms to fear stimulates the creation of exosomes to detoxify this fear. Scientists have mistakenly labeled these tiny bodies "viruses"—poisons. They are not pathogenic viruses; they are the natural human response to fear, lies, hatred, and other toxins. Exosomes are nature's way of letting us know that unless we rid ourselves of these toxic thoughts, healthy life is not possible. Masks, social distancing, closing businesses, violence, and racial intolerance are just some of the forms of fear-inducing strategies to which humans are subjected. Human beings need love, trust, and acceptance to grow and thrive. These waveforms are out there as well. Our challenge is to learn how to tune into these good emotions rather than the things that bring illness and death.

PART 3

CHOICES

CHAPTER 12

QUESTIONING COVID

The first case of illness attributed to a "contagious" organism called coronavirus occurred in China in November 2019; the illness appeared in the United States in January 2020. By the end of June, halfway through the year, health officials could cite ten million cases with a half a million deaths worldwide.[1] The official numbers in the United States are about 2.5 million cases with 126,000 deaths,[2] or a death rate of about 5 percent. The official cause: person-to-person transmission of a "novel" form of a type of virus called coronavirus, an organism listed in textbooks as causing mild upper respiratory symptoms or the common cold.[3] Officials and the media have studiously avoided mentioning any possible connection with the stealthy installation of 5G antennas, first in large cities and then in smaller towns. These antennas are deliberately inconspicuous, hardly noticeable on city streets.

Examples of 5G Small Cell Installations

A large portion of these deaths (43 percent or higher) have occurred in nursing homes or long-term-care facilities.[4] The elderly are the most vulnerable, with an average age at death of seventy-nine years. Nearly all victims have comorbidities such as obesity, diabetes, high blood pressure, and heart disease, which means they are probably taking several toxic drugs, like metformin for diabetes, ACE-2 inhibitors for high blood pressure, and statin drugs to lower cholesterol. According to Silvio Brusaferro, president of the Italian Higher Institute, Italian medical records indicate that "there may be only two people who died from coronavirus in Italy, who did not present other pathologies."[5]

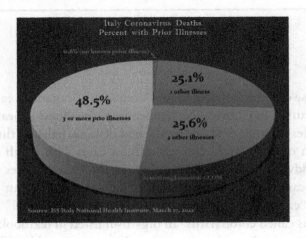

In the early days of the illness, many argued that the threat of coronavirus was exaggerated. In March, Stanford Professor John Ioannides claimed that health officials were severely overreacting to coronavirus, suggesting that "the response to the coronavirus pandemic may be 'a fiasco in the making' because we are making seismic decisions based on 'utterly unreliable' data."[6]

A report from March 9, 2020, shows deaths attributed to Covid-19 at fifty-six per day, versus malaria at two thousand per day and TB at three thousand per day[7]—hardly a mortality rate that justifies pandemic status, especially as physicians report receiving pressure to write Covid as the cause of death on death certificates.[8] Hospitals have ample reason to list Covid as a cause for admission; they receive $13,000 from Medicare when they list a patient labeled as "Covid" compared to only $4,600 for simple pneumonia. If the patient is put on a ventilator, Medicare pays the

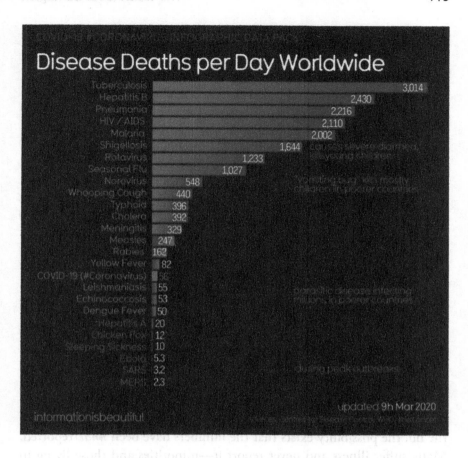

Disease Deaths per Day Worldwide

Tuberculosis	3,014
Hepatitis B	2,430
Pneumonia	2,216
HIV / AIDS	2,110
Malaria	2,002
Shigellosis	1,644
Rotavirus	1,233
Seasonal Flu	1,027
Norovirus	548
Whooping Cough	440
Typhoid	396
Cholera	392
Meningitis	329
Measles	247
Rabies	162
Yellow Fever	82
COVID-19 (#Coronavirus)	56
Leishmaniasis	55
Echinococcosis	53
Dengue Fever	50
Hepatitis A	20
Chicken Pox	12
Sleeping Sickness	10
Ebola	5.3
SARS	3.2
MERS	2.3

updated 9h Mar 2020

informationisbeautiful

hospital $39,000.[9] These financial incentives made it easy to argue that the Covid-19 incident and mortality rates were inflated.

Early analyses of the US death rate claimed a "virtually nonexistent" increase in the number of US deaths during the first seventeen weeks of 2020 compared with the same period in 2019. But using more complete CDC data reveals that over a twelve-week period (February through April), Covid-19 claimed more lives than did accidents, stroke, diabetes, suicide, and other conditions. Covid-19 was third in the list of leading causes of death in the United States during the twelve-week period. The overall US death rate was 4–5 percent higher than it was in the same period in 2019.[10]

Some have claimed that Covid deaths are mostly iatrogenic—that is, caused by the medical care patients receive and by the many toxic drugs they are taking.[11] Typically, a Covid-19 patient receives the antiviral remdesivir and is put on a ventilator. Like AZT for AIDS patients, remdesivir was developed to treat another disease—hepatitis C, for which it did not work as hoped—and dusted off to give to Covid-19 patients. Adverse effects include respiratory failure and organ impairment, low albumin, low potassium, low red blood cell count, low platelet count, gastrointestinal distress, elevated liver enzymes, and reaction at the site of injection.[12]

In the early days of the pandemic, the media reported a rush to produce enough ventilators to satisfy the expected demand for them. But the much-hyped ventilator turned out to be a death sentence. According to one analysis, among patients sixty-six years and older hospitalized in the New York City region, those put on ventilators had a 97.2 percent mortality rate.[13] In an April 22, 2020, article published in the *Journal of the American Medical Association*, an analysis of 5,700 Covid-19 patients hospitalized between March 1 and April 4 found that the overall death rate was 21 percent, but it rose to 88 percent for those who received mechanical ventilation.[14]

Bad medical care was not the only factor that contributed to the high death rate; the other was almost certainly terror and loneliness. When tagged with a diagnosis of Covid-19—either from a putative positive test result or no test at all—patients often found themselves locked up against their wills in elder care facilities and shut off from the outside world—no visits from family or friends permitted.

Although many argue that the rates of illness blamed on the coronavirus have been inflated, and that Covid-19 is no worse than a bad case of the flu, the possibility exists that the numbers have been *under*reported. Many suffer illness and never report it—minorities and those living in poverty, but also those who mistrust the medical system; and we have no idea how many are really getting sick in China or in countries formerly of the Communist world. The PCR tests give false positives, but also false negatives, which means that many may suffer a mild form of the illness without a diagnosis.

More important, it is clear that symptoms of Covid-19 are *not* the same as those of the ordinary flu. Autopsy reports have found that the lungs of Covid victims contain microscopic blood clots—something that does not happen in flu patients. In larger blood vessels of the lungs, the number of blood clots is similar among Covid-19 and flu patients. However, capillaries in Covid-19 patients have nine times more blood clots than do those in flu victims. The capillaries are in the small air sacs that allow oxygen to pass into the bloodstream and carbon dioxide to move out. Actually, pathologists are finding clots in almost every organ.[15]

The damage, of course, is blamed on the wily virus: "The novel coronavirus is a master of disguise . . . [and] . . . uses a number of tools to infect our cells and replicate."[16]

According to Professor Mauro Giacca of King's College London, Covid-19 often leaves lungs completely unrecognizable. "What you find in the lungs of people who have stayed with the disease for more than a month before dying is something completely different from normal pneumonia, influenza or the SARS virus. You see massive thrombosis. There is a complete disruption of the lung architecture—in some lights you can't even distinguish that it used to be a lung."

"There are large numbers of very big fused cells which are virus positive with as many as 10, 15 nuclei," he reported. "I am convinced this explains the unique pathology of Covid-19. This is not a disease caused by a virus which kills cells, which has profound implications for therapy."[17] The "viruses" of course are exosomes trying to remove toxins from the lung cells; but they are apparently no match for serious EMR poisoning, which seems to completely disrupt the structure of lung cells.

A key symptom of Covid-19 is prolonged and progressive hypoxia—meaning that the body is starved for oxygen. This happens when the hemoglobin molecule releases its iron molecule. Unattached iron in the bloodstream is reactive and toxic, but normally iron is tucked away in the hemoglobin molecule—the iron is caged, so to speak, and carried around safely by hemoglobin. (Vitamin C has an important role to play in cleaning up rampaging iron ions.)

Without the iron ion, hemoglobin can no longer bind to oxygen, so cannot carry oxygen to the cells. Meanwhile, the released iron does its reactive damage everywhere in the body. Damage to the lungs shows up in CT scans. The kidneys release hormones like erythropoietin, which tell your bone marrow to ramp up production of hemoglobin.

The conventional explanation for the release of iron from hemoglobin is the action of glycoproteins in the coronavirus—but the action of 5G's millimeter waves is an equally good explanation, especially those at 60 GHz, which disrupt oxygen molecules. An interesting observation about lung malfunction in Covid-19 patients is that it is bilateral (both lungs at the same time), whereas ordinary pneumonia typically affects only one lung.[18] What kind of virus knows to attack both lungs?

A study from Wuhan showed that more than one-third of coronavirus patients had neurologic symptoms including dizziness, headaches, impaired consciousness, skeletal-muscle injury, and loss of smell and taste—and more rarely seizures and stroke.[19] This is not your normal flu, this is a serious disease.

Moreover, in late March, reports of Covid-19 deaths in infants began to appear.[20] In the early months, the disease mostly afflicted the elderly, but doctors are observing an increase in an inflammatory system called Kawasaki disease, which afflicts children and teenagers. Called "pediatric multi-system inflammatory syndrome temporally associated with Covid-19," it is diagnosed on the basis of symptoms. These symptoms include high fever, rash on trunk and groin, extremely red eyes, dry cracked red lips and a strawberry-red swollen tongue, redness and extensive peeling of the hands and feet, and swollen lymph nodes. Severe abdominal pain and gastrointestinal symptoms, inflammation of the heart muscle, and markers of cardiac injury are other typical symptoms of Kawasaki disease.[21]

However, ironically, the overall death rate among children has declined during the pandemic lockdown, from seven hundred deaths per week to well under five hundred by mid-April and throughout May, a change attributed to parents not keeping their children up with draconian vaccination schedules.[22]

Since remdesivir gave disappointing results, health officials are seeking other remedies. One suggestion is dexamethasone, a potent steroid that can shrink the brain. Dexamethasone makes sense if Covid-19 is an inflammation rather than an "infection."[23] In fact, one of the first things medical students learn is that steroids like dexamethasone make infections worse. Since dexamethasone may make Covid-19 better, this demonstrates that the illness can't be an infection.

Another proposed treatment is the drug Haldol (haloperidol), sometimes called vitamin H.[24] Haldol is one of the most potent antipsychotic medications in existence—it puts the patient into a kind of drooling stupor. Doctors and scientists from France report serious effects when nonsteroidal anti-inflammatory drugs (NSAIDs) such as ibuprofen are given to Covid-19 patients.[25] NSAIDS can cause internal bleeding, as do ACE inhibitor blood pressure medications.

Official policy states that no coronavirus drug is universally safe and effective and discourages nontoxic or holistic treatments, but the lack of successful treatments from mainstream medicine has patients seeking alternatives. A report at the end of March brought attention to the work of Dr. Vladimir Zelenko, a New York physician who claims to have treated almost seven hundred coronavirus patients with 100 percent success using a malaria drug called hydroxychloroquine sulfate with supplemental zinc, a treatment that costs only twenty dollars over a period of five days.[26] Success is probably due to the patient receiving zinc and sulfur. A study published on May 22 with great media attention in both *The Lancet* and the *New England Journal of Medicine* claimed that the treatment was useless and warned that it could possibly cause death. But on June 4, *The Lancet* retracted the study and apologized to its readers. "The study was withdrawn because the company that provided data would not provide full access to the information for a third-party peer review . . . Based on this development, we can no longer vouch for the veracity of the primary data sources."[27]

Dr. David Brownstein reports no hospitalizations in eighty-five patients diagnosed with either Covid or suspected Covid using vitamins A, C, and D, hydrogen peroxide, and iodine, while advising patients to avoid the flu vaccine.[28]

Ozone is another promising therapy.[29] Other proposed alternative treatments include acetazolamide (altitude sickness drug), hydrogen peroxide IV, vitamin C IV, hyperbaric oxygen, hydrogen gas, and chlorine dioxide,[30] but none of these are available in hospitals.

After Memorial Day, Covid-19 hot spots flared in Arizona, Oklahoma, South Carolina, and Florida, which officials blamed on relaxed mitigation efforts—having fun at the beach or visiting bars, not wearing face masks, and not practicing social distancing. According to one official, "There are certain counties where a majority of the people who are tested positive in that county are under the age of thirty, and this typically results from people going to bars."[31] Are these new cases simply due to increased testing with many false positives?" Or to the wily virus infecting people with person-to-person transmission? Or to the continued deployment of 5G technology to smaller cities and to the Southwest, and longer exposure to 5G as the weeks and months go by?

In mid-June, government agencies could point to an increase in cases in Texas, Alabama, and Virginia. "The findings indicate that the risk for large second waves of outbreaks remains low if communities continue to implement cautious, incremental planned re-openings that limit crowding and travel to non-essential businesses . . . without vigilance in masking, hygiene and disinfection, certain southern counties will remain high risk."[32] Health officials warned that opening states too soon could have "disastrous consequences." In early July, Texas reversed course and mandated masks.

Although increased testing with the useless PCR test has undoubtedly generated higher numbers of Covid cases, hospitalizations have also increased. The curve has not flattened, it's going up again.[33] Hospitalizations have also gone up in California, in spite of strict masking and social distancing measures there since early in the year.

Sweden initially appeared as a bright spot among nations by forgoing a mandatory lockdown, with factories, businesses, bars, and restaurants remaining open, and a lower illness and death rate than other European nations. Whereas tourism came to a halt in the rest of Europe, it flourished in Sweden. However, case numbers and deaths began to rise in April, with the total number of deaths now over five thousand. Was this because Sweden failed to follow lockdown and mask mandates? Or was it the rollout of 5G, beginning in March? An April 6 article stated: "Sweden is in the process of introducing super-fast 5G mobile telecoms networks, giving users several times faster web speeds compared with existing 4G technology.[34] Sweden's first Covid-19 death was March 10.[35]

The fact that no explosion of cases occurred in large cities after the Memorial Day protests has puzzled health officials. Areas of unrest like New York, Chicago, Minneapolis, and Washington, DC did not see any increase in cases even though thousands of protesters did not wear masks nor practice social distancing. Of the thirteen cities involved, only Phoenix saw an increase in cases and hospitalizations, which officials blamed on a decision to end Arizona's stay-at-home order and ease restrictions on businesses: "Arizona residents who were cooped up for six weeks flooded Phoenix-area bar districts, ignoring social distancing guidelines."[36] The wily coronavirus apparently zeroed in on these law-abiding citizens, but not on protesters crowding the streets.

The official government policies for curtailing Covid-19 are self-isolation, social distancing, hand washing, surface cleaning (environmental hygiene), and face masks—because "Coronavirus can spread by just talking or breathing." The only recommended treatment for those in an acute stage of infection is ventilation.

Many have pointed out that the pores in even the best facemasks (even the N95 respirators) are ten times larger than any "virus." A study published in May 2020 in *Emerging Infectious Diseases* reviewed the evidence for the effectiveness of "nonpharmaceutical personal protective measures and environmental hygiene measures in nonhealthcare settings." The evidence from fourteen randomized controlled trials of these measures did not find that hand washing, environmental hygiene, or use of face masks had any effect on reducing transmission of so-called infectious diseases.[37] What's more, labels on boxes of masks specifically warn that the masks "will not provide any protection against COVID-19 (Coronavirus) or other viruses or contaminants."

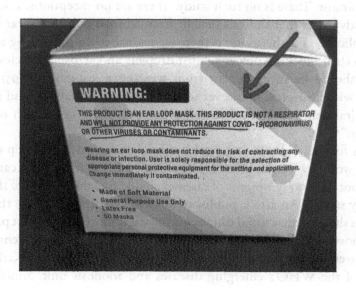

Moreover, wearing a mask can cause serious adverse health effects, including headaches, increased airway resistance, carbon dioxide accumulation, and hypoxia, especially the tight-fitting N95 respirator mask.[38] In one study, one-third of health-care workers wearing the N95 respirator mask developed headaches and 60 percent required pain medications for relief.[39] Wearing a mask can cause a reduction in blood oxygenation (hypoxia) or an elevation in blood CO_2 (hypercapnia). The N95 mask, if worn for hours, can reduce blood oxygenation as much as 20 percent, which can lead to loss of consciousness. The Occupational Safety and Health Administration (OSHA) warns that masks risk creating a low oxygen environment and are not effective in preventing disease.[40]

Recently two boys in China fell dead wearing masks while running during physical education classes.[41] Nevertheless, Los Angeles officials have decreed that Angelinos must wear a face covering while outdoors. The new law requires face coverings for walking, running, cycling, scooting, roller skating, skateboarding, and all outdoor activities except for those on the water.[42]

A recent review of the science relevant to Covid-19 social policy concluded that masks and respirators don't work. "No [randomized control trial] study with verified outcome shows a benefit for [healthcare workers] or community members in households to wearing a mask or respirator. There is no such study. There are no exceptions. Likewise, no study exists that shows a benefit from a broad policy to wear masks in public. . . . Furthermore, if there were any benefit to wearing a mask due to their blocking power against droplets and aerosol particles, then there should be more benefit from wearing a respirator (N95) compared with a surgical mask, yet several large meta-analyses, and all the RCT [randomized control trials], prove that there is no such relative benefit."[43]

As for social distancing, health officials have put a guilt trip on the whole world by warning that asymptomatic carriers (those who carry the virus but have no symptoms of disease) could "fuel the spread" of the disease by stealth. An article published at GreenMedInfo.com lists thirteen studies showing that social distancing increases mortality in heart patients and those with diabetes, causes depression, and generally shortens life.[44] Moreover, such policies do no good. In June, Dr. Maria Van Kerkhove, head of the WHO's emerging diseases and zoonosis unit, announced, "From the data we have, it still seems to be rare that an asymptomatic person actually transmits onward to a secondary individual."[45]

As for environmental sanitation, some of the guidelines border on the ridiculous. One article suggests that flushing the toilet without a lid could spread coronavirus.[46] The theory is that through flushing one's stool down the toilet it gets aerosolized into the air. Obviously, you can't transfer the virus to yourself so it's not a big deal if you watch the contents of your toilet get flushed away. And members of your household have already been exposed to your "viruses," so the real risk is for those who invite groups of strangers to watch their stool get flushed down the toilet. Citizens are warned to refrain from this dangerous habit.

Even though the science does not support social distancing and the use of masks as a way to control disease, school officials are seriously proposing masks and social distancing for grade schoolers when they return to the classroom in September. The school district of New Albany, Ohio, has taken such policies even further. In addition to masks and social distancing, the school district would require each student to wear an electronic beacon to track their location to within a few feet throughout the day. The device will record where students sit in each classroom, show to whom they meet and talk, and reveal how they gather in groups. These devices could also be used on younger students who do not have smartphones.[47] To compile all the data from the tracking beacons, the schools will need the kind of powerful Wi-Fi service that only 5G—mounted inside the buildings—can supply.

Without the virus theory—and even *with* the virus theory—masks, social distancing, and lockdown make no sense. Ironically, the advent of 5G and EMFs from cell phones and other devices gives us good reason to avoid crowded situations. Take choir practice, with a few dozen people in close quarters, most with cell phones in their pockets and possibly a cell phone tower in the church steeple: this is the perfect situation for creating illness in electrically sensitive individuals.

Think meatpacking plants with hundreds of people standing at close quarters, all with cell phones, possibly with 5G installed inside the building to track product, and the added electro-smog stress of conveyer belts and the constant whirring electrical machinery. (Workers in small plants, typically located in more rural areas and lacking conveyer belts and other electrical equipment, will be less vulnerable.)

Think schools, office buildings, universities, and stadiums—where 5G is planned, and indeed has been installed covertly (under the guise of disinfecting) during the coronavirus lockdown. The new wave of illness predicted in September is likely, with students going back to classrooms, newly wired for 5G (and again, most with cellphones).[48] Or imagine tens of thousands crowded into a stadium, now fitted out with 5G so everyone can use their phones. "Verizon is building its 5G Ultra Wideband network to support transformational changes across multiple industries including sports and entertainment," said Kyle Malady, Verizon's chief technology officer. "This next generation technology can enhance the fan experience with the potential to revolutionize the way sports are viewed and played. Faster download speeds, higher bandwidth and lower latency on 5G-enabled mobile devices with Verizon's 5G Ultra Wideband service is just the beginning."[49]

Outbreaks of illness in factories, schools, theaters, and stadiums are sure to happen, and will appear to confirm the contagious virus theory.

Health officials who have predicted a "second wave" in September will have the satisfaction of being right . . . and the justification to proceed with their promised solution: a vaccine.

CHAPTER 13

A VACCINE FOR COVID-19

The story of vaccination begins in 1796 with Edward Jenner of Gloucestershire, who administered the first vaccine against smallpox. At that time, many country folk observed that dairymaids generally had beautiful unmarked skin, proof they had never contracted smallpox. The reason, some claimed, was that their exposure to cowpox, said to be a mild bovine variation of smallpox, gave them immunity to the human form of the disease. Some country folk believed the superstition that after a case of cowpox, you could never get smallpox—a belief contradicted by the observations of physicians at the time.[1]

Of course, there is a far better explanation for the beautiful skin of milkmaids—unlike most people of the time, they had daily access to a superb source of nutrition (including a rich source of vitamin C) and nature's premier probiotic food. Good nutrition (including good probiotics) protected milkmaids against diseases like smallpox—good nutrition protects all of us from disease. Of course, preventing bedbugs with good sanitation also plays an important role, but it is unlikely that milkmaids slept in clean beds. If they did get bitten, the superb nutrition of their milk-based diet protected them.

On May 14, 1796, Jenner tested his hypothesis by inoculating James Phipps, an eight-year-old boy who was the son of Jenner's gardener. He scraped pus from cowpox blisters on the hands of a milkmaid and scraped the pus into the arm of the child.

Jenner excelled at self-promotion, and in 1802, the English government awarded him ten thousand pounds for continued "experimentation."

Jenner claimed that his vaccine gave perfect immunity for life. Unfortunately, statistics taken from *Reports of the Registrar General of England* indicate that the vaccine was not a success, with deaths from vaccination outnumbering deaths from smallpox up through the early 1900s.[2] In 1831, a smallpox outbreak in Wurtemberg, Germany, claimed the lives of almost one thousand people who had received a vaccination; and in the same year, two thousand vaccinated people in Marseilles, France developed smallpox. In 1854–1863, following the introduction of compulsory vaccination programs in Europe, smallpox claimed over thirty-three thousand lives, and other epidemics followed, epidemics in which thousands of vaccinated people died. Compulsory vaccination laws in England were repealed in 1907, by that time their failure too obvious to disguise.

During one of the worst smallpox epidemics in England, between 1870 and 1872, the city of Leicester took a different approach. They instituted a system of good sanitation and quarantine, with the result that there was only one death from smallpox in Leicester during the epidemic.

Louis Pasteur followed in Jenner's footsteps with the development of a vaccine for chicken cholera (which didn't work) and for anthrax (which was forced on many farmers, who reported that their sheep died anyway[3]) and finally, in 1855, for rabies (also called hydrophobia). The premise was that if you could create a less virulent source of disease and inoculate that into a healthy person, the person would develop immunity and not get sick when he encountered the full-blown illness.

Even in Pasteur's day, physicians doubted that hydrophobia was a specific disease; dogs became vicious through starvation and neglect; and neurological disease leading to insanity could happen in the aftermath of any kind of wound, especially a puncture wound. The most likely cause of rabies is a form of tetanus or botulism—both are associated with clostridium toxins that the bacteria produce under anerobic conditions, as in puncture wounds. Doctors in Pasteur's day had excellent results treating dog bites by cauterizing them with carbolic acid. One doctor reported cauterization of about four hundred dog bite victims without one developing a case of hydrophobia.[4]

But Pasteur believed that he could prevent rabies by vaccinating the victims of dog bites. He created the rabies vaccine by taking saliva, blood, and part of the brain or spinal cord (usually the cerebrospinal fluid) from a suspected animal and injecting it into a living rabbit, then aging and drying the cells from the rabbit's spinal cord so that it could be injected into human beings.

His first patient, a badly bitten nine-year-old boy, received the vaccine—*after a doctor had cauterized the wound*—and recovered. Pasteur proclaimed his success—but others were not so lucky. A Dr. Charles Bell Taylor, writing in a publication called *National Review* in July 1890, listed many cases in which Pasteur's patients died, whereas the dogs that had bitten them remained healthy.[5] One of these cases was King Alexander of Greece. In his own reports, Pasteur fudged the numbers to make it seem as though he had a high rate of success.[6]

In the practice of medicine, sometimes the dogma about a certain subject becomes so rigid that even when the truth is sitting in plain sight, physicians simply can't see it. Such is the case when it comes to the concept of "permanent" immunity to infectious disease. Medical students are taught early in their training that our immune systems are organized around the principle that if we get an infectious disease once, we will never get it again. This is supposedly due to the two phases of our immune system working together to create memory of a virus or bacteria that lasts for life. Scientists have worked out the intimate details of this immune memory over the last century in order to provide the theoretical basis for vaccines. The immune theory also derives from the simple observation that nobody gets the typical childhood diseases twice in their lives. Like many ideas in medicine, however, the truth may be far more complicated.

Two simple observations cast doubt on the premise of immunity for life. The first is that certain bacterial diseases, such as strep throat, do tend to recur; in fact, there is no immunity-for-life construct when it comes to bacterial infections. As for viral infections, we can easily observe that most people get numerous colds and flu throughout their lifetimes. Scientists usually explain this by saying these are just different "viruses" one catches, and just because you are immune to one, doesn't confer immunity to other viruses.

We also know and generally accept the fact that a child who gets chicken pox is susceptible later in life to a second manifestation of chicken pox called *shingles*. Shingles is believed to be caused by the same virus, but with a different symptom picture. Similarly, many people have experienced repeated bouts of cold sores or "herpes" outbreaks. So, there is no lifelong immunity in these cases.

Interestingly, at the time that Pasteur and others were formulating the germ theory of disease, along with its corollary of immunity for life, many scientists and physicians disagreed. One of the dissenters was Professor Alfred Russel Wallace. In his book *The Wonderful Century*, he had the following to say about smallpox:

> Very few people suffer from any special accident twice—a shipwreck, or railway or coach accident, or a house on fire: yet one of these accidents does not confer immunity against it happening a second time. The taking it for granted that second attacks of smallpox, or of any other zymotic disease, are of that degree of rarity as to prove some immunity or protection, indicates the incapacity for dealing with what is a purely statistical question.

Wallace describes a study by Dr. Adolf Vogt, professor of Hygiene and Sanitary Statistics at the University of Berne, Switzerland. Vogt compiled data on those who had contracted smallpox and their subsequent susceptibility to further smallpox episodes. He found that those who had one episode of smallpox were actually 63 percent more likely to suffer from a second episode as compared with those who had never had a case of smallpox. Vogt concluded: "All this justifies our maintaining that the theory of immunity by a previous attack of smallpox, whether the natural disease or the disease produced artificially, must be relegated to the realm of fiction."

Wallace went on to prove that people vaccinated against smallpox actually had a much higher death rate from smallpox than did those who were not vaccinated. In particular, Wallace studied the high death rate from smallpox of the heavily vaccinated soldiers in the US military compared with the results obtained from the sanitation methods employed by the town on Leicester in England. Here is what Wallace concluded:

> It is thus completely demonstrated that all the statements by which the public has been gulled for so many years as to the almost complete immunity of the revaccinated Army and Navy are absolutely false. It is all what the Americans call "bluff." There is no immunity. They have no protection. When exposed to infection they do suffer just as much as other populations or even more. In the whole of the nineteen years 1878–1896 inclusive, unvaccinated Leicester had so few smallpox deaths that the Registrar-General presents the average by the decimal 0.01 per thousand

population equal to ten per million, whereas during 1876–1889, there was less than one death per annum. Here we have real immunity, real protection; and it is obtained by attending to sanitation and isolation, coupled with the almost total neglect of vaccination. Neither Army nor Navy can show any such results as this.[7]

Clearly, protection through vaccination for diseases like smallpox is an ugly superstition that must be discarded. At the same time, we must abandon the concept of lifelong immunity conferred by the activity of our immune system.

But what about the observation that children essentially never get measles twice in their lives? In this case, there is so little actual study of this phenomenon that it is difficult to make any firm conclusions. But we must remember that the typical childhood diseases such as measles, mumps, whooping cough, and chicken pox are best understood as processes of normal growth and maturation for the child. If this is the case, there would be no reason for children to go through these processes more than once in their lives. After all, a tadpole transforms into a frog only once; a caterpillar becomes a butterfly only once.

Measles is a process of detoxification, transformation, and growth. If thwarted, particularly by an injection that clearly alters our "immune" responses, then only worse things can happen. This is clearly revealed by the many studies showing that children who go through typical childhood illnesses such as measles have fewer chronic diseases throughout the course of their lives. The body likely makes a chemical or protein that we call an antibody to mark this event. But it is far from clear that antibodies are protective of anything, or that these childhood diseases are contagious. We must have the courage and insight to rethink this whole concept of disease.

For the pharmaceutical companies, however, the concept of introducing a small amount of a bacteria, virus, or toxin into the body to create lifelong immunity supports the practice of vaccination, and by the early twentieth century, they were having a field day producing vaccines for any disease they could think of. American soldiers made convenient Guinea pigs and complained in letters home of receiving a vaccination every week. Many have surmised that the effects of the Spanish flu were exacerbated on military bases by all these vaccinations inflicted on the troops, including a crude bacterial meningitis vaccination experiment.[8] (Another factor that resulted in the high death toll among American soldiers was the use of aspirin, often in huge doses, which undoubtedly contributed to the excessive hemorrhaging that carried so many away.)

The process for creating modern vaccines involves many trade secrets and numerous ingredients. Unknown to the public is ongoing controversy over which process produces the best and safest results—because serious reactions to vaccinations are commonplace. When one understands the basics of how all modern viral vaccines are manufactured, however, it becomes immediately clear not only how fraudulent the whole process is but also how the production of modern viral vaccines helps prove that these viruses can't possibly be causing the diseases for which they are blamed.

To produce a modern vaccine, technicians first collect biological fluids from an infected person, usually respiratory secretions or the fluid from skin lesions. This presumably contains millions of copies of the virus—along with an untold number of components from cellular debris. The fluid is then centrifuged to concentrate the virus. The next step is to inoculate this centrifuged fluid onto a variety of tissue cultures, usually tissue derived from monkey kidney cells, aborted human fetal tissue, or chicken eggs. Some companies have proposed using cancer tissue as a culture, due to the fact that cancerous tissue is "easier to grow in large amounts," but this practice is still considered too risky.

It turns out the viruses that supposedly will kill us all aren't strong enough to infect the tissue cultures. This means in order to help the virus taken from the sick patient lyse (kill) the tissue cells, they have to starve and poison the tissue first. Once the tissue is weak enough, the virus can then infect the cells, inject its genetic material into the cells, and produce millions of copies of itself. That's the theory, anyway.

The resultant infected tissue is an unholy mixture of the original snot (now frozen and distributed across the globe to all the various vaccine manufacturing companies as their "stock" material), the toxins (antibiotics, oxidizing agents, etc.) used to weaken the tissue, the debris from the breakdown of the tissue, and the "viruses" that emerge from this process. There are often some light purification steps added at this point, but never anything even remotely approaching isolation and purification of the viruses. Then finally, some preservatives (usually mercury for multi-dose vials, still used for the flu vaccine) and stabilizers (such as polysorbate 80, an emulsifier that breaks down the blood-brain barrier) are added to this final product. This is a live viral vaccine.

A "dead" or attenuated viral vaccine is all of the above steps and then a final heat or chemical sterilization step at the end to "kill" or at least neuter the virus. Never mind that one can't say viruses are even alive in any meaningful sense of that word, nevertheless, they are "killed," usually with heat, in this step. Then an adjuvant, usually aluminum, is added to the final product to make sure that the person receiving the mixture attempts to eliminate it from the body and so produce antibodies, considered proof of an immune response that will protect against the disease.

Another way to make an attenuated or inactive viral vaccine is to start with dead, killed tissue. Then technicians isolate single protein particles from the dead tissue. Sometimes they even produce these particles synthetically. After that they add adjuvants and preservatives, including aluminum. Aluminum added to the rest of these toxic chemicals is the likely culprit for the creation of the excessive inflammatory reactions that frequently occur with any attenuated viral vaccine—and are one of the central hallmarks of the Covid-19 syndrome. It would certainly be an interesting research project for someone to track the relationship between prior vaccine usage and subsequent development of Covid-19 symptoms in adults and children.

The message that Americans get every morning in the newspapers and every evening on the news is that a Covid-19 vaccine will save us—with mandatory vaccines for everyone, we can go back to life as normal, and scientists will have saved the day.

The vaccine would be liability-free and rushed to market, making its debut in January 2021. Liability-free means that the consumer has no redress, no matter how bad the injury to himself or his child, no matter how costly the care after the injury; and it means that vaccine manufacturers have absolutely no incentive to make a vaccine that is either safe or effective.

Even if a virus is the cause of Covid-19, manufacturers face a number of obstacles. For one, the virus already "has mutated into at least thirty different genetic variants."[9] The variants include nineteen never seen before as well as "rare changes that scientists had never imagined could happen." In addition, the challenge of producing so many vaccines in so short a time is daunting.

As described in an article published by Children's Health Defense,[10] the solution proposed by pharmaceutical companies is a new type of vaccine that can "outsmart" nature using next-generation vaccine technologies such as gene transfer and self-assembling nanoparticles—along with invasive new vaccine delivery and record-keeping mechanisms like smartphone-readable quantum dot tattoos—which will require the vast capacity of 5G networks to read and process.

To quickly produce a vaccine for the whole world, they will also need to develop new manufacturing techniques that circumvent the slow processes of traditional vaccine production. The new techniques use genetic engineering (recombinant DNA technology) subjected to "expression systems" (bacteria, yeast, insect cells, mammalian cells, or plants such as tobacco)—to produce so-called "subunit vaccines." The problematic hepatitis B vaccine was the first to employ this entirely new vaccine production approach, and a number of the Covid-19 vaccines currently under development are deploying these techniques. However, subunit vaccines must be bundled with "immune-potentiating" adjuvants (most likely aluminum) that can trigger an inflammatory immune response.

Even newer are DNA and messenger RNA (mRNA) vaccines, which are basically a form of gene therapy. Whereas traditional vaccines introduce a vaccine antigen to produce an immune response (which doesn't actually mean that the recipient is immune), nucleic acid vaccines instead send the body instructions to produce the antigen itself. As one researcher explains, the nucleic acids "cause the cells to make pieces of the virus," so that the immune system then "mounts a response to those pieces of the virus."

DNA vaccines are intended to penetrate all the way into a cell's nucleus. According to one biotech scientist, "This is an incredibly difficult task given that our nuclei have evolved to prevent any foreign DNA from entering."[11] Maybe nature has a reason for protecting the nucleus from genetic invasion!

When some DNA vaccines made it into clinical trials in the late 2000s, they were plagued by "suboptimal potency," meaning they didn't work. Scientists then came up with the idea of augmenting vaccine delivery with "electroporation"—electric shocks applied to the vaccine site (using a smart device) to make cell membranes more permeable and force the DNA into the cells. Electroporation remains a key design feature of some Covid-19 vaccine candidates today.

A second aspect of DNA vaccines—their gene-altering properties—is also troubling. DNA vaccines, by definition, come with the risk of "integration of exogenous DNA into the host genome, which may cause severe mutagenesis and induced new diseases." Framed in more understandable terms, "disruption from DNA is like inserting a foreign ingredient in an existing recipe, which can change the resulting dish." The permanent incorporation of synthetic genes into the recipient's DNA essentially produces a genetically modified human being, with unknown long-term effects.

Regarding DNA gene therapy, one researcher has stated, "Genetic integrations using viral gene therapies . . . can have a devastating effect if the integration was placed in the wrong spot in [the] genome." Discussing DNA vaccines specifically, the *Harvard College Global Health Review* notes that the DNA vaccines could cause chronic inflammation, because the vaccine continuously stimulates the immune system to produce antibodies. Other concerns include the possible integration of foreign DNA into the body's host genome, resulting in mutations, problems with DNA replication, autoimmune responses and activation of cancer-causing genes—think children with birth defects and cancer early in life.

The mRNA vaccines are "particularly suited to speedy development" and have attracted attention as the "coronavirus frontrunners." mRNA vaccines can reportedly generate savings of "months or years to standardize and ramp up . . . mass production." mRNA vaccines need reach only the cell cytoplasm rather than the nucleus—an apparently "simpler technical challenge"—although the approach still demands "delivery technologies that can ensure stabilization of mRNA under physiological conditions." This involves "chemical modifications to stabilize the mRNA" and liquid nanoparticles to "package it into an injectable form."

Unfortunately for the pharmaceutical companies, mRNA vaccines have displayed an "intrinsic" inflammatory component that makes it difficult to establish an "acceptable risk/benefit profile." mRNA enthusiasts admit that there is, as yet, an inadequate understanding of the inflammation and autoimmune reactions that may result from the vaccine. This raises the specter of a true disaster should regulators grant the manufacturers of Covid-19 mRNA vaccines their wish for "a fast-track process to get mRNA vaccines to people sooner."

A good example of a rushed vaccine was the dengue vaccine experiment—which actually increased the risks of dengue fever:[12] Dengue fever is a common disease in more than 120 countries and, like coronavirus, has been the target for a vaccine for many years. The development and licensure of Dengvaxia® vaccine by Sanofi spanned more than twenty years and cost more than 1.5 *billion* US dollars. But the development of the vaccine turned out to be difficult. Dengue vaccine antibodies often made

the infection worse—called "disease enhancement" in vaccine-speak—
especially in infants and children. When the vaccine was administered
to thousands of children in the Philippines, at least six hundred died.
The Philippine government has permanently banned the vaccine from
the country.

Another rushed vaccine, for the swine flu, was a total fiasco. Early in
1976, after several soldiers became severely ill at Fort Dix in New Jersey,
supposedly from swine flu, President Gerald Ford announced a plan
to rush through a vaccine so that every American could be vaccinated.
But manufacturers balked at the specter of liability for vaccine injuries
and one company produced two million doses with the "wrong strain."
Congress passed a law waiving liability, and Ford pushed plans to inoculate one million people per day by the fall, even though reports seeped
through that the vaccine had caused injuries and was not effective. In
mid-October, Ford went on television to show himself receiving an injection from a White House doctor. Meanwhile, the capricious swine flu
failed to appear and in December, following ninety-four reports of paralysis from the vaccination, the program was terminated, and the danger of
swine flu disappeared from the pages of the newspapers.[13]

So far, trials for the coronavirus vaccine have not gone well. On
May 18, 2020, Moderna Inc. (co-owned by the National Institutes of
Health[14]), headquartered in Cambridge, Massachusetts, announced that
it had obtained "positive interim clinical data" from a Phase I clinical
trial for an mRNA Covid vaccine. Moderna stock soared (and their top
executives sold off over one hundred million dollars' worth of shares).[15]
However, four of the forty-five participants suffered serious reactions.
The three subjects who received the highest doses all experienced grade-
three systemic symptoms, which can mean blistering open ulcers, wet
peeling, or serious rash over large areas of the body. The press release did
not mention the results of other trials.[16]

One volunteer, Ian Haydon, age twenty-nine, stated that the vaccine left him "the sickest he's ever been." He was rushed to urgent care,
where he almost fainted. But Haydon is still cautiously optimistic about
a vaccine.[17]

The New York Times reported positive results from a vaccine in development by the University of Oxford. "Monkeys given an experimental
vaccine from the University of Oxford appear to have resisted the novel
coronavirus. Six rhesus macaques given hAdOx1 nCoV-19 in Montana
did not fall ill despite heavy exposure," was the headline.[18] But they did

fall ill, in fact, all the vaccinated macaques sickened after exposure to Covid-19, "suggesting the treatment, which has already received in the region of £90 million in government investment, may not halt the spread of the deadly disease."[19]

An experimental vaccine for Covid-19 that uses human fetal cell lines, in development by CanSino Biologics, Inc. of Tianjin, China, also had poor results. In a clinical trial involving 108 volunteers, ranging in age from forty-five to sixty years old, 81 percent suffered at least one adverse reaction within seven days after vaccination. Adverse effects included fever, fatigue, headache, and muscle pain, some of it severe.[20]

Following these disappointing results, the FDA relaxed the rules. On June 30, the agency announced that any Covid-19 vaccine would have to prevent disease, or decrease its severity, in only 50 percent of the people who receive it.[21] The American College of Obstetricians and Gynecologists (AGOC) has suggested testing Covid-19 experimental vaccines on pregnant women.[22] And, the officials tell us, the vaccine might be needed multiple times, perhaps annually.[23]

It's clear that a vaccine is not going to save us—in fact it has the potential of inflicting enormous suffering on the world's population, not to mention violent resistance to the idea of universal gene modification by electroporation. And all for an illness that is not contagious!

Only two things are going to resolve the coronavirus problem. One is a new system of etiquette. Just a few decades ago, few people gave much thought about smoking in someone else's house; today such an action is considered the height of rudeness. Today, no one with any sense would light up in front of another person without asking permission, and certainly would not smoke in another person's dwelling place. Today we are shocked to see men and women smoking in old movies—we all know that tobacco companies paid producers to show glamorous people smoking cigarettes, and we shake our heads in disgust.

In the future, we will feel the same disgust when we see people in the movies hold cell phones to their ears. How could filmmakers encourage such a dangerous practice! Like the warnings on cigarette packages, there will be warnings on cell phones against use by children; sales of cell phones to young people will be forbidden. And no one will dream of entering another person's home with their cell phones on. Public pressure will ensure that all large gatherings—sports events, concerts, fairs, conventions, choir practice, dress rehearsals, church services, and private parties—begin with admonitions to turn cell phones to airplane mode.

High schools will not allow cell phones on campus and all computer input for children will be in wires. Offices will designate special areas for cell phone use, in deference to the electrically sensitive, like special areas for smoking, and give all employees old-fashioned telephones.

The second task involves a massive cleanup. Just as the Industrial Revolution created unsanitary conditions that took decades of patient hard work (as well as new technologies) to amend, the wireless revolution will require the same patient remediation—mainly by putting as much communication as possible into wires, but also by exploring new technologies for mitigating EMFs in the home and office. New houses will be built with minimizing EMF exposure in mind, and old houses will be upgraded—just as old houses eventually got bathrooms and central heating. These measures are not as glamorous as introducing a vaccine that will bring fame and fortune to a few, but they are the only real solution to the electro-smog pollution of the Internet age.

And there is some good news. For years the telecommunications companies have collected fees from your phone bill in order to bring high-speed fiber optics to every home, school, and business in America—fees amounting to more than five thousand dollars per household and totaling billions of dollars. But instead of putting fiber optics everywhere—the job is only about 50 percent complete—these companies invested this money illegally to force people into wireless plans. Accounting tricks have made fiber optics services appear unprofitable while wireless seemed extremely profitable. The telecommunications companies, in collusion with the Federal Communications Commission (FCC), have used these distorted financial results to argue that they cannot bring wired Internet to rural areas or even inner cities. More important, these accounting tricks have provided an excuse to shut off the wired networks and go wireless with 5G. Some telecommunications executives have even proposed getting rid of old-fashioned telephone service.

Fortunately—most fortunately—a recent court case *IRREGU-LATORS v. FCC: DC Court of Appeals Opinion*, March 13, 2020, removes FCC jurisdiction and returns it to state regulatory agencies. All illegal subsidies for wireless can now be stopped—and 5G is no longer profitable.[24]

CHAPTER 14

5G AND THE FUTURE OF HUMANITY

"What a piece of work is man!" In Hamlet's famous soliloquy, Shakespeare poses questions about the nature of the human being. One of history's greatest initiates and thinkers, Shakespeare describes the human being as the crowning achievement of creation, free and unlimited in potential. Humankind is noble, created in the image of the Godhead itself, yet subject to all the foibles, temptations, and errors with which we are all well acquainted.

The question for this chapter, having previously explored the watery nature of life itself, is what or who is this human being? The corollary to the question, one that is rarely asked, is why we should even care whether human beings will survive this current pandemic or whether we are about to go extinct, as have so many fellow animal species? In some ways, if we can't begin to answer this simple but profound question, then what difference does it make whether or not there are pathogenic viruses creating contagions? To answer this question—what is a human being?—is actually the key to bringing the Covid-19 phenomenon to a successful resolution; it is the key to the challenge of building a world we all know is possible, but may be too afraid to pursue or even demand.

When one tries to pinpoint the unique characteristics of human beings, many answers come forward. Some will say that this is a stupid or unanswerable question. Others might suggest that human beings are "apes without hair," "the only upright animal," or "the animal with the largest brain." The scientist might point to our unique genetic makeup, the only living being with this particular set of forty-six chromosomes. A religious person might say that human beings are created in the image of God and as such are given dominion over the earth.

The technologist might deny anything particularly special about human beings and point out our many design flaws, which he believes could be improved or upgraded by merging human beings with computers, creating a kind of download for our minds. Unfortunately for them—fortunately for human beings—the technocrats have been unable to locate the human mind.

These various descriptions all have certain elements of truth in them—well, maybe not the downloaded mind theory—but they all miss one simple and clear distinction, one that is unarguable and scientifically irrefutable. It is the one that holds the key to understanding Covid-19.

The difference between the human being and all the rest of the animals is simple: the human being is the *only* living being that can bear children at every single time of the year. All other animals go into heat, and become fertile only at specific times of the year, usually so they give birth in the spring—wild animals go into heat once a year and domestic animals (dogs, pigs, cows, etc.) typically twice a year, and a few (cats and rabbits) several times per year. But human beings, at least at this point in our evolution, are able to conceive at any time of the year. People on earth are conceived every single day of the year and people are born every single day of the year. No animal can do this. What is the significance of this obvious fact?

As we have discussed in chapter 7, all matter is simply the congruence of wide and varied "forces," or "energies," which make up our entire universe, received or collected through the universal receptor we call water. The recognition of these universal energies or forces was the basis of all traditional wisdom and healing systems until the current materialistic medical model came to the fore, and under which we all now labor.

Traditional healers recognized the influence of the stars and planets; for example, they characterized chamomile as a "Venus" plant, stinging nettles as a "Mars" plant, and dandelion as a "Jupiter" plant. Now that we understand the electromagnetic nature of the universe, these characterizations begin to make sense.

This foundational understanding was the basis of most human knowledge until it was lost in recent times. Although this way of looking at the world is key to our development as individuals, it needs to be rediscovered because the deployment of 5G technology threatens the foundation of existence, and we can understand why this is so only if we recapture the ancient way of thinking about life.

Unlike animals, the human being is not conceived or born under the influence of any specific cosmic energy field. Instead, each human being is conceived and born at a unique time and place, therefore under the influence of a unique cosmic field. This is the physiological basis of our freedom and individuality. This is what makes the human being the crown of creation; this is the basis for the claim that the human being is created in the image of God, God being a concept embracing the energetic field of the entire universe. Each individual human being is a unique component in this field, and humanity as a whole is the sum of the entire field and therefore the image or the reflection of God. This is the essential message of all major philosophical and religious traditions from the past.

These cosmic fields in the form of electromagnetic forces radiate toward the earth from all parts of the cosmos and are "collected" in the ionosphere or electromagnetic shield of the earth. In a way similar to how we obtain nourishment from our food, these electromagnetic forces are "processed" by this protective skin of the earth, the ionosphere. When we take in food, the food is processed by the bacteria, fungi, viruses, and other microbes in our gut; their "waste" then becomes the nourishing food we absorb to give us life. Similarly, the cosmic forces are worked on in our ionosphere, to emerge as the life-giving electromagnetic fields that nourish the earth and all life, including human life. That is how it has always worked and how it should work—and it would except for the introduction of the technology we call 5G.

Without getting too deep into the engineering of 5G technology, the important thing about 5G is that the pulsed millimeter waves, this new "spectrum" that will run our phones and computers faster, need to be "organized" by placing hundreds of thousands of satellites right into the ionosphere of the earth. These hundreds of thousands of satellites will emit their own electromagnetic frequencies that essentially beam these new, man-made signals down to the millions of receivers placed in our neighborhoods, stadiums, schools, nursing homes, hospitals, parks, farms, lakes, forests, oceans, and everywhere else on earth. The intention is to blanket the earth with these manmade electromagnetic fields.[1] We know that these millimeter waves interfere with the availability of oxygen in the atmosphere and hence will also interfere with the ability of the

mitochondria (bacteria) in our tissues to convert oxygen into energy. This is the main feature of 5G, exacerbated by aluminum poisoning, glyphosate poisoning, general air pollution, and all the many other toxins in our modern world—all contributing to the symptoms of "Covid-19."

But none of this can match the consequences of putting hundreds of thousands of satellites into the ionosphere of the earth. If this is allowed to happen, not only will all life on earth be subject to the constant toxic effect of these harmful millimeter waves, as if that isn't bad enough, but the direct consequence of this folly is that the cosmic waves that come to us from the furthest reaches of our cosmos will no longer be allowed to maintain their integrity in their journey to the earth. Life will be cut off from the cosmos, elk will no longer be elk, squirrels will have lost the energetic source that makes them squirrels, and human beings will no longer be formed as free individuals with their own unique destinies. Life will become formed under the influence of computer code written by the self-styled new "masters of our universe." And, all this so we can have faster download speeds for our cell phones.

Humanity is at a crossroads, and although we can present mitigation strategies that transform the energy fields that constitute 5G technology (see Appendix B), we should be clear. "Covid-19" is the first wave of disease created by the introduction of this new technology. It is only the tip of the iceberg. Officials warn us that more waves are coming. They know. They are replacing the wisdom of God with the folly of man. It's time for humanity to wake up, grow up, and to find the courage to stop this menace.

EPILOGUE

Once upon a time, in a far-off place, lived a king and a queen. Their kingdom was happy, prosperous, and peaceful. Unfortunately, the royal couple was barren, and the people were worried that they would be left with no suitable heir to the throne.

One day, the king and queen were walking in the forest and they came upon a pond. Out of the pond jumped a frog who informed them that they would soon have a beautiful daughter. Naturally, this news filled them and the entire kingdom with joy.

To celebrate the arrival of their daughter the royal family invited the twelve wise women of the kingdom to a feast at the palace. After the feast, each of the wise women spoke up and conferred a blessing on the new child. One gave the blessing of beauty, another gave the blessing of kindness, another of wisdom, and so on to include all the good qualities that ennoble the human being. After the eleventh wise woman had spoken, an uninvited woman burst into the palace banquet hall. She was also an elder woman, but unlike the others she had evil intentions. She was a witch, and she was angry that she had not been invited to the banquet to honor the new child.

Out of her fury, she pronounced a terrible curse on the child, saying that as the child grew into adulthood, she would prick her finger on a spindle and fall down dead. The people in the hall were stunned and shocked. Thankfully, the twelfth wise woman had not yet spoken. Upon hearing this curse, she announced that the evil woman was powerful, and therefore she could not completely undo this curse, but she could change it. Instead of falling down dead, the young girl, if she pricked her finger on a spindle, would only go to sleep, not die.

After the banquet, the king ordered the destruction of all the spindles in the kingdom. There would be no possibility that the princess could ever prick her finger on a spindle.

As time went on, the princess grew into a beautiful young woman, with all the wonderful traits bestowed upon her by the wise women. She was kind, beautiful, and wise, and the entire kingdom prospered.

One day the king and queen left the palace for a state outing. As usual, the princess was left in the charge of state officials, whose orders were to keep a close eye on her. Everyone loved the princess, and being of a curious nature she convinced the officials to let her wander freely in the castle. She came upon a room she had never seen before. Inside was an old woman sitting on a stool spinning wool. Curious, having never seen anyone spin wool before, she asked the old woman what she was doing.

The old woman handed her the spindle; the princess pricked her finger and as predicted fell into a deep sleep. When the king and queen returned to the castle, they put her in a beautiful bed. Then everyone else in the kingdom fell asleep as well. The bread was still in the oven, the shoemaker's leather still on the bench, the shepherds' flocks were still in the fields. The entire kingdom fell under the influence of an evil spell.

As time went on, the castle became engulfed in a mass of thorns and vines. Anyone visiting the kingdom from another land was unable to penetrate the deep thicket of toxic thorns. Those who tried were met with certain death. So it was for one hundred years.

One day, a young prince from a far-off land was out hunting and came upon the overgrown castle. An old man told him about the plight of the beautiful young princess inside. Something moved him deeply; he knew he could and must save the princess and the kingdom. He announced his intentions to his mother and father, the king and queen, and to the entire court. They forbade him to undertake this quest, knowing he would meet certain death.

He responded simply, "I am not afraid, I will rescue the beautiful princess."

And, so it came to pass. Without any fear in his heart the thorns had no power over the prince; as he approached the castle, they parted before him. He found the princess lying on her bed; he was astonished by her radiant beauty. He knelt and kissed her. The princess awoke, and with her the entire kingdom; everyone simply resumed their business. The prince and princess were married, the kingdom was restored to health, prosperity and peace, and they all lived happily ever after.

Most of you will recognize this story of Sleeping Beauty or Briar Rose. It is a story told to children all over the world for centuries, to impress on their souls the ways of the world and to give them courage. It is also the story of "Covid-19" if we know how to properly unravel its metaphors.

The world was peaceful and prosperous but it was clear that without a true heir—without a true direction—the world would not be able to go on as it was. Then, as if out of nowhere, a lowly frog—a representation of the spiritual world in fairy tales—informs the king and queen that there is a road to continued prosperity and fruitfulness of the land—a princess shall be born. Overjoyed, the royal couple invite the twelve wise women to celebrate this news. The twelve wise women represent the collected wisdom of the universe. They are the zodiac, the cycle of the year, and the connection between the cosmos (the entire zodiac) and the earth (the cycle of the year).

But there is a thirteenth woman, the evil witch. Witches are typically depicted in fairy tales as having distorted, twisted bodies, yellow eyes, and sallow skin. They are the picture of illness, the incarnation of the materialistic view of life. They are fallen matter, hence some sort of crone, stepmother, or weak mother figure. The wise women are the spiritual view of life, the witches or fallen women are the materialistic view of life.

The twelfth wise woman cannot undo the power of the witch's curse, it is too powerful; in fact, the curse is something the kingdom *must* go through to achieve health, prosperity, and peace on a higher and lasting level. She can, however, mitigate its power. And so it is. The entire world goes into lockdown. Anyone trying to rescue the world, anyone who points out that life doesn't have to be like this, is torn asunder by the toxic thorns that engulf the kingdom. The entire world is cursed, living as if in a dream or spell, the curse of materialism in all its guises, for the promise of faster video games. Just like us.

But there is a way out, a way discovered by the prince from a far-off land. He expresses this clearly: "I am not afraid." Without fear, guided by love, the curse can be broken. The world can be restored, the lesson can be learned, materialism, currently in the form of a toxic virus theory and enthrallment to the Internet, can be overcome. We can do this, it's just a spell, it's not real, it's make-believe. All we have to do is find courage and love in our hearts to embrace the truth. That is all that matters.

APPENDIX A

WATER

As described in chapter 8, water that heals, the kind of water that the healthiest human cultures have all consumed, shares a number of specific characteristics. To reiterate, healthy water is free of *all* toxins, it contains the full complement of minerals, and it is structured through subjection to a constant vortex motion. As a result of this constant vortex motion, the water is oxygenated. The oxygenation step is crucial in producing healing waters, and is a characteristic of the most renowned healing waters on the planet, like the water at Lourdes and in the Ganges River. Finally, the water is "finished" through its exposure to nature sounds and frequencies all along its path. When these steps are followed, we have a healing water that is one of the core foundations of health for plants, animals, and humans. Providing healing, healthy water like this should be a basic requirement for every hospital, hospice, and healing institution.

Interestingly, in addition to the water we drink, it is clear from both historical evidence and modern research that the water we wash and bathe in is at least as important as the water we drink. This point was brought home to us as a result of personal experience and looking at the research on the Ophora water systems. Ophora is an innovative water company located in Southern California, which has developed a technique for taking any municipal or well water and cleansing it of *all* traces of toxins, pharmaceuticals, fluoride, chlorine, and even microplastics. As far as we know, no other system is able to purify water to this level. Then the water is re-mineralized by adding all the known minerals found in seawater. After that, the water is put through a rose quartz vortex and oxygenated to forty parts per million with a proprietary technology. This is by far the highest oxygen content of any water tested, on a level with the rarest and finest healing waters on earth. The water is pH balanced and finished by exposing the water to the sounds of nature.

Studies of people who soak in this oxygenated water and drink one-half gallon per day show some amazing results. First, the oxygen saturation of the tissues immediately goes up and stays at the highest levels for up to twelve hours. This is a more robust oxygen saturation response than one gets with hyperbaric oxygen therapy, a strategy known to produce many therapeutic effects.

Second, and even more astonishingly, the phase angle (PhA) measurement of the tissues also rises.[1] The phase angle measurement tells us about the level of hydration of the tissues and is actually an indication of the subject's biological age.[2] PhA is a direct measurement of your cell membrane integrity.[3] The membrane is what structures the water inside and outside of your cells. The membrane is where the cells communicate with one another and where an electrical charge is created so the cell can function. (See chapter 9 for a discussion of the appropriate diet for cell membrane integrity.)

The phase angle measurement goes up within hours of a one-hour soak in a bath of oxygenated, purified water along with drinking one-half gallon of the water over the course of a day. Furthermore, this improvement in the phase angle, a process that usually takes months of detoxification and the cleanest of diets, seems to last for some time, after even just one session of soaking and hydration. One can only guess at the improvement in the quality of hydration and the biological age of the person if this became a weekly or daily practice. Imagine the results that hospitals would achieve if they began treatment with oxygenated drinking and bathing water, rather than putting patients on ventilators!

For purity and oxygenation levels, the Ophora water system is currently the only system that we can recommend for creating therapeutic waters. This presents a dilemma as most readers will find purchasing either an Ophora system for home use or actually purchasing the water directly from Ophora (Ophorawater.com) to be cost prohibitive. Clearly the best solution would be for every town and municipality to use multiple Ophora water systems to produce healthy water for its citizens. The Ophora owners are ready and prepared to help with this project.

For personal showering and bathing, the simplest options are either an Ophora shower device, which eliminates some toxins from the water while oxygenating and structuring it, in conjunction with some kind of filter or reverse osmosis system in your house. Another possibility is an Aquadea showerhead attachment for your shower or bathtub. The Aquadea system puts the incoming water through a high-speed implosion vortex so that it essentially sucks the water out of the pipe rather than pushing it as is otherwise the case. (This is exactly how the heart helps move the blood, through suction rather than "pushing."*)

The suction is easily confirmed by putting one's hand over the water as it emerges from the showerhead. Rather than your hand being pushed, you will feel your hand sucked up against the showerhead. The vortex created by the arrangement of crystals inside the showerhead creates a high-speed implosion effect, an effect that imbues the water with energy and life.

An interesting experiment is to water one group of plants with Aquadea water and the second group with normal water. Showering under an Aquadea showerhead is like taking a shower under a waterfall and bathing in Aquadea water is like bathing in a flowing stream—the invigorating effects are similar. The Aquadea people can customize the type of crystals they use in producing your showerhead and the materials (usually gold, silver, or bronze) that the showerhead is made of. The drtomcowan.com website is currently the US distributor for Aquadea showerheads. There are a variety of more affordable options for home filtration systems that could be combined with an Aquadea showerhead.

A good option for home-delivered water is Mountain Valley Spring (mountainvalleyspring.com), which has delivery service nationwide and uses glass bottles only. Castle Rock (castlerockwatercompany.com) bottles water in glass and is available in stores nationwide.

A simple way to mineralize and oxygenate these recommended waters is to fill an eight-ounce glass, add a pinch of unrefined sea salt, and stir with a long-handled spoon, making a vortex. Stir in one direction and then reverse to stir in the other direction, repeating several times.

APPENDIX B

BIO-GEOMETRY AND EMF MITIGATION

The interesting and amazing thing about the current state of human awareness is that what we generally call science is actually mostly a series of easily disproven, irrational superstitions. Here is one simple example of such a superstition, which once corrected holds the key to protecting ourselves and all of nature from the harmful effects of EMFs. Start by asking yourself or your friends the simple question: does the form, shape, and pattern of some object, and the quality of the materials that make this form, shape, or pattern, have an effect on the invisible energies of living things? Science, at least medicine, is clear that such an idea is unscientific nonsense. If a doctor at a conventional medical conference suggested that one could lay a geometric form on a human being or place a geometric form near a human being to produce a therapeutic effect, he would be a laughingstock.

However, consider the Stradivarius violin, widely considered the best violin ever made, some of them selling for tens of millions of dollars. What is this violin? Simply put, it is a specific geometric form made from a specific material called *moon wood* (timber harvested during the waning moon when the sap in the trees is at its lowest) that somehow forms invisible sound waves into unmatched music. The sound this specific form, made with this specific material, produces has been prized for centuries by violinists all over the world. Apparently, the form and material of this specific violin is able to shape invisible sound waves to create the most exquisite music. Only a completely unscientific mind would conclude that this is the only example in nature in which form, patterns, and shape made out of a specific material affects the energy that we now know animates life.

The reality is that this process of form, shape, pattern, and material affecting the energy around us is the norm and not some unusual occurrence. It is the basis of "sacred" science all over the world, even back to the time humans lived in caves and carved geometric forms into the walls of the caves. From our earliest days, through the forms of the pyramids, monuments, and sculpture, humans have been working with material and form to create effects on the life around them. It is only the modern, irrational materialistic "scientists," whose ideas are formed primarily by superstition, who deny these effects are real.

Bio-geometry is simply the science of how form, shape, and patterns combined with certain materials shape the energy in the world around us. The result of this shaping is either a beneficial or harmful effect on all life forms including human beings. Living in a time of increasing electro-smog, with no apparent end in sight for the level of pollution we will be exposed to, it is imperative that all our readers explore the strategies and techniques offered by bio-geometry to mitigate these effects. This is not in any way to suggest that putting 5G satellites and towers up is somehow fine if we just use bio-geometry to mitigate their effects—this is not at all true. These insane installations must be stopped. At the same time, right now, everyone can and should avail themselves of the science of bio-geometry to deal with the current electromagnetic pollution we all face.

No disrespect is meant to the many companies that are using waves, patterns, forms and other techniques to mitigate EMF effects. Some of them are helpful. However, based on studies with plants, animals, and humans, as well as our personal experience, the strategies used by bio-geometry stand above all the rest in terms of their safety and efficacy.

For example, in the late 1990s, the National Liver Disease Research project in Egypt undertook a study of patients with hepatitis C and elevated liver enzymes. Although most claim that hep C is a viral disease, the elevation of the liver enzymes does not argue for the presence of a virus but indicates some toxin that is affecting the health of the liver. In this study the participants were asked to wear either a bio-geometry pendant, use a placebo, or follow conventional antiviral therapy. The researchers followed the course of the liver enzymes in the six months after the intervention. The head of the study, Dr. Tasha Khalid, announced on Saudi television that the results showed 90 percent of the participants who used the bio-geometry pendants had a reduction in liver enzymes in the first six months. This is in contrast to a reduction of 50 percent in those who used conventional treatment and 20–30 percent of those in the placebo group.[1] This is a dramatically positive result for such a simple and inexpensive intervention, and one that should be followed up with other studies on other illnesses.

The best way to use bio-geometry is to become a bio-geometry practitioner yourself. You can do this by contacting the Vesica Institute and signing up for their online classes (vesica.org). The next best thing is to contact and work with a trained bio-geometry practitioner who can get you started in mitigating your personal space, including helping you find personal protective devices to use. The final approach, one that is the least everyone should use, is to purchase the L90 and bio-emitter pendants from the vesica.org website and use them as directed all the time.

The vesica.org website also features the fascinating work of Ibrahim Karim, who was able to mitigate the effects of radio antennas in a church steeple in Hemberg, Switzerland, using various shapes strategically placed in the church and nearby houses.[2]

In addition, there are many commonsense precautions that everyone can take. If you use a cell phone, keep it to a minimum and never put it to your ear. Have a landline in your home for regular telephone use. If possible, use wired Internet, not Wi-Fi. At least turn off your Wi-Fi at night. Don't let your utility company install a smart meter; there is often a fee to pay for keeping an old-fashioned analog meter, but it is worth it.

While you sleep, your bedroom should be free of electro-smog. Turn off the fuses to all wiring in your bedroom at night; you can even have an electrician install a switch to do this near your bed. Don't have an electric alarm clock on your nightstand, near your head. Use a battery-powered alarm clock instead.

Avoid electric cars, fluorescent and compact lights, "smart" appliances, and electric-powered exercise equipment.

Above all, spend some time each day outside, far away from antennas—in a park, on a farm, or in the woods—to give your body the rest it needs from EMF pollution.

APPENDIX C

WHAT TO EAT?

A healthy diet, one that helps maintain your cellular gels, gives you maximum energy, and protects you as much as possible from EMFs, does not require any renunciation. You don't have to eat a diet that is dry and tasteless, but can enjoy a variety of delicious, satisfying food. What is required is an attention to quality and preparation methods. In other words, we have to think carefully about every morsel of food we put in our mouths. In traditional cultures, they ate the food they had and knew instinctively how to prepare it; they did not have to think about how to have a healthy diet, it just happened naturally.

The mark of modern man is that he is an individual, and he can no longer depend on his tribe or village to make decisions for him. He is faced with a bewildering variety of choices, especially in the food he eats. He must wade through the myriad offerings of processed food (much of it addictive) and also be wary of misinformation, especially the misinformation coming from the medical establishment; he must avoid falling for a diet of industrial foodstuffs while also steering clear of weird, invented diets and especially low-fat diets that are impossible to follow.

In terms of quality, avoid industrially farmed food. Someday everyone will "have a farm"—meaning that everyone will know a particular grass-based farmer from whom they purchase their meat, poultry, eggs, and dairy foods. There is no way that healthy animal foods can be raised in an industrial system. To find these foods, visit realmilk.com, or contact your nearest local chapter of the Weston A. Price Foundation (westonaprice.org).

For grains and produce, buying organic is important—especially when it comes to wheat products. Buying organic bread, crackers, and flour will ensure that the wheat has not been sprayed with glyphosate just before harvest. Organic produce is now widely available, even in supermarkets.

An important resource is the *Shopping Guide* published by the Weston A. Price Foundation. Updated yearly, the guide names over 1,600 products categorized as "best" or "good." The guide is free to members and can also be purchased at westonaprice.org. For recipes, see *Nourishing Traditions: The Cookbook that Challenges Politically Correct Nutrition and the Diet Dictocrats.*

GET YOUR FATS RIGHT

First and foremost, we need to get our fats right, because the right fats help maintain our cell membranes and contribute to stable structured water in our tissues; the right kind of fats carry critical vitamins that sustain and protect every system in the body.

Instead of margarines and spreads	Use butter instead
Instead of cooking oils	Cook in lard or bacon fat
Instead of commercial salad dressings	Make your own with olive oil and vinegar
Instead of Cool Whip	Use real whipped cream
Instead of nondairy creamer	Use real cream or real half-and-half
Instead of commercial mayonnaise	Make your own or use a coconut oil–based mayo
Instead of commercial dips	Make your own using sour cream and other ingredients
Instead of chips	Crunch on plain pork cracklings
Instead of typical crackers	Find crackers made with palm oil or coconut oil
Instead of processed snacks	Enjoy natural cheese and artisan salami
Instead of supermarket bread	Use the WAPF Shopping Guide to find natural sourdough bread without added oils
Instead of French fries	Make your own oven fries cooked in lard or duck fat

Instead of fast-food fried chicken	Make your own, fried in lard
Pastries, cake, donuts	Minimize as best you can (drink a glass of raw milk instead!)
Pizza	Save for a special treat, not every day, and order thin-crust pizza

SWEETS

The bane of the modern diet, refined sweeteners should be minimized—we know that is hard! Refined sweeteners include sugar, high-fructose corn syrup, agave syrup, glucose and fructose. You will find that eating the right fats will help reduce your cravings. And you don't have to deprive yourself entirely. Homemade desserts made with natural sweeteners are fine in moderation. Eat sweet foods after a meal so that they don't cause a blood sugar roller coaster!

Instead of refined sweeteners	Use natural sweeteners like maple syrup, maple sugar, raw unfiltered honey, and rapadura (dehydrated cane sugar juice)
Instead of commercial baked goods	Make your own, using real ingredients like eggs, butter, nuts, and natural sweeteners
Instead of soft drinks	Enjoy kombucha, kefir, and other lacto-fermented beverages (now widely available in supermarkets). Be careful to choose those with a low sugar content
Instead of candy	Snack on real food like cheese, artisan salami, nuts, and fresh fruit
Instead of commercial ice cream	Make your own ice cream with real cream, egg yolks, and a natural sweetener

GRAINS

Instead of most commercial bread	Make the effort to obtain genuine sourdough bread made with organic grains
Instead of extruded breakfast cereals	Make cooked porridge, soaked overnight

| Instead of most commercial crackers | Choose crackers listed in the Weston A. Price Foundation shopping guide |

SAUCES & SEASONINGS

Instead of commercial salt	Use unrefined salt, as much as you like
Instead of commercial sauces high in MSG	Make your own sauces and gravies using genuine bone broth
Instead of commercial seasoning mixes	Use real herbs and organic spices

DAIRY PRODUCTS

Instead of pasteurized & ultra-pasteurized milk	Enjoy whole raw milk from pastured cows
Instead of processed cheese	Eat real cheese, preferably from raw milk
Instead of commercial, sweetened yogurt	Make your own or use plain whole yogurt
Instead of industrially produced butter	Purchase grass-fed butter

EGGS

| Instead of commercial eggs | Purchase your eggs from a farmer who raises the hens in the out of doors on pasture |
| Instead of egg whites only | Use the whole egg, even with added yolks |

MEAT

Instead of industrially raised meat	Purchase grass-fed meat from a farmer
Instead of commercial meat products	Purchase artisan salami, ham, bacon, etc.
Instead of muscle meats only	Eat liver and other organ meats as pâté, terrines, scrapple, liverwurst, etc.

SEAFOOD

| Instead of farmed fish and shellfish | Enjoy wild-caught fish and shellfish |

SUPPLEMENTS

Instead of fish oil	Use natural cod liver oil (listed in the Weston A. Price Foundation shopping guide)
Instead of industrial vitamin C	Use products that are powders of vitamin C—rich foods
Instead of synthetic vitamins	Use desiccated nutrient-dense foods like desiccated heart, liver, oysters, etc.

FERMENTED FOODS

| Instead of modern pickles | Use raw sauerkraut and other lacto-fermented foods |

Addendum

Since publication of *The Contagion Myth*, a number of new studies have appeared or have come to our attention. These shed new light on the points we made in the book, proving that there is simply no scientific case to be made for the existence of a SARS CoV-2 virus, nor any justification to the claim that it causes illness. Many of these papers are difficult for a lay person to interpret; many contain statements that once interpreted are shocking to read.

In this addendum we provide the relevant citation, quote the most important passages from the article and then give a short explanation for the significance of this finding. As always, we don't claim to be the sole keepers of the truth, but we do insist on a proper scientific debate based on facts, not propaganda. In that spirit, we encourage other scientists to comment on the facts of our conclusions, rather than on our right to say them.

1. Corman VM and others, "Detection of 2019 novel coronavirus (2019-nCoV) by real-time RT-PCR.," *Euro Surveill*, Jan, 2020; 25(3):2000045.

> The ongoing outbreak of the recently emerged novel coronavirus (2019-nCoV) poses a challenge to public health laboratories as virus isolates are unavailable.

The authors—more than twenty virologists who work in public health—were tasked with developing tests to accurately determine who is infected with the new virus and who is not. As they say in the introduction, there are no virus isolates available, which actually means they have no idea whether the virus actually exists. Furthermore, there is no possible way to devise a test for something you haven't isolated or analyzed. The situation is similar to asking someone to devise a test for unicorns while forbidding them from finding out whether unicorns actually exist. No wonder these scientists find this task "challenging!"

Further on, they reiterate the problem: "We aimed to develop and deploy robust diagnostic methodology for use in public health laboratory settings without having virus material available."

Read that again: their job is to develop a test to detect the new virus but they don't have any samples of the new virus to work with. This is nothing short of shocking to read.

2. Harcourt J and others, "Severe Acute Respiratory Syndrome Coronavirus 2 from Patient with Coronavirus Disease, United States," *CDC Bulletin* **Volume 26, Number 6, June 2020.**

In the Results section, the authors state:

> Therefore, we examined the capacity of SARS-CoV-2 to infect and repli-
> cate in several common primate and human cell lines, including human
> adenocarcinoma cells (A549), human liver cells (HUH7.0), and human
> embryonic kidney cells (HEK-293T) . . . Each cell line was inoculated at
> high multiplicity of infection and examined 24h post infection. No CPE
> was observed in any of the cell lines except in Vero cells.

Note, CPE means "cytopathic effect," which refers to structural changes in host cells that are caused by "viral invasion." The infecting virus is said to cause lysis (breaking up) of the host cell or, when the cell dies without lysis, an inability to reproduce. Both of these effects are said to occur due to CPEs.

While this may not mean much to most readers, it is in reality a shocking statement. Virologist have three "hosts" they can use in their attempts to prove that viruses cause illness. After "isolating" the virus, they can expose humans to the virus; they can expose animals to the virus; or they can use tissue cultures taken from various animal or human sources and expose the tissue culture to the virus. Leaving aside the fact that they never actually isolate and purify the virus, which they openly admit, let's assume that the unpurified fluid they are using does contain the relevant virus and therefore should be able to transmit infection.

In the history of virology, most virologists have decided not to do their experiments on human subjects, as this is considered unethical. In the case of the SARS-CoV-2 virus, we know of no published study that used humans as the test subjects.

Virologists also admit that in the case of most viral infections, there are no studies available that prove infection in animals. How a virus can infect and kill humans but not animals is left unexplained. Researchers get around this obvious biological conundrum by saying, "there are no animal models on which to test such-and-such a virus." In other words, "We know that the virus infects and kills humans even though we've never tested the virus on humans because that would be unethical. Therefore, we do our tests on animals, even though when we test animals they don't get sick, because they are not proper 'hosts' for the virus. So, you'll just have to trust us."

In the case of SARS CoV-2, there are two studies we know of that used unpurified "virus" on animal models, one with hamsters and one with mice. In the hamster study,[1] researchers took the unpurified, lung-cancer-grown, centrifuged animal secretions, and squirted it down the throats and into the lungs of a group of unfortunate hamsters. Some, but not all, of the hamsters got pneumonia, and some even died. We have no idea what would have happened if they had squirted plain lung cancer cells into the lungs of these hamsters, but probably not anything good. And even more perplexing, some of the hamsters didn't even get sick at all, which certainly doesn't square with the deadly contagious virus theory.

In the mouse study,[2] researchers infected both transgenic mice and wild (normal) mice to unpurified virus. None of the wild mice exposed to the "virus" got sick. Of the mice genetically programmed to get sick, a statistically insignificant number either lost some fur luster or had an insignificant weight loss. Thus, scientists have not been able to show that the Covid-19 "virus" causes harm to animals.

At least 95 percent of virological proof-of-contagion studies are carried out in animal or human tissue. In other words, virologists never expose humans to a virus, and they rarely expose animals. Instead, to prove infection, they grow tissues taken from various sources and expose the virus to these tissues. Never mind that in order for the "virus" to infect any tissue culture, they have to starve the tissues first in a "minimal-nutrient medium" and then poison the tissues with gentamicin and amphotericin, two potent antibiotics that are nephrotoxic—toxic to the kidneys.

Let's see what they got from exposing human tissue cultures to the "virus." As you can read, in all three of the human cell lines no CPE (no cell death, no infection) was observed. Only Vero cells (monkey kidney cells) were adversely affected—and remember, the material injected into these cells contained kidney toxins. So basically they proved that the SARS-CoV-2 virus does *not* infect human tissue. Interestingly, in their conclusion the authors don't mention this important fact. Only virologists reading the whole paper will find out that if you want to grow the virus, don't bother to use human cell lines.

3. CDC Document, "Novel Corona virus (2019-nCoV) Real-Time RT-PCR Diagnostic Panel," July 13, 2020.

On page thirty-nine of the document, under Performance Characteristics we read:

> Since no quantified virus isolates of the 2019-nCoV are currently available, assays (diagnostic tests) designed for detection of the 2019-nCoV RNA were tested with characterized stocks of in vitro transcribed full length RNA....

Thus, as recently as July 2020, the CDC is publicly admitting—in writing—that they have no "isolates" of the 2019-nCoV-2 virus. Instead, they are basing their analyses and test preparations on "manufactured" pieces of RNA, which have no proven relationship to any virus. Here we have clear language from the CDC saying the virus has never been seen, isolated, or characterized.

How anyone thinks they can make a test to detect something that has never been isolated and characterized strains any sense of credulity or scientific reasoning. This is a clear and damming statement showing the complete lack of evidence for SARS CoV-2. And, something imaginary cannot possibly cause disease except through psychological methods. There is no viral disease called Covid-19

4. Yuxiang W and others, "RNase-Resistant Virus-Like Particles Containing Long Chimeric RNA Sequences Produced by Two-Plasmid Coexpression System," *Journal of Clinical Microbiology*, 2007, DOI: 10.1128/JCM.02248.

We find the following quote under Figure 3 of the study:

> Lane 1, positive control of pET-MS2-3V plasmid using the primers S-SARS1 and HA300RT-A; lane 2, RT-PCR of SARS-Cov! Plus SARS-CoV 2 plus SARS-Cov3 plus HCV+HA300...

This paper is by a group of Chinese virologists who were testing and reporting on a new and different technique for characterizing viruses—nothing particularly funny about that. The funny part is they are claiming to be using primers (genetic sequences) taken from various SARS viruses. They include the SARS–CoV-1, CoV-2, and CoV-3 viruses in their analysis. Wait a minute, a SARS-CoV-3 virus? We didn't know there was a SARS-CoV-3 virus, what are they talking about?

To make this paper even more interesting, note the date of the publication: 2007. In other words, this paper is claiming to have sequences isolated from the SARS CoV-2, called a *novel* coronavirus—and this is 2007! We have heard of something called time travel, but this finding makes no sense at all.

Endnotes

Preface

1   Robert Williams, *Toward the Conquest of Beriberi* (Cambridge, MA: Harvard University Press, 1961), 18.
2   MJ Rosenau, "Experiments to Determine Mode of Spread of Influenza," *Journal of the American Medical Association* 73, no. 5 (August 2, 1919): 311–313.
3   "Cells and viruses vocabulary," https://quizlet.com/171172750/cells-and-viruses-vocabulary-flash-cards/.
4   CD Bethel et al, "A National and State Profile of Leading Health Problems and Health Care Quality for US Children; Key Insurance Disparities and Across State Variations," *Academic Pediatrics* 11, no. 3S (May-June 2011).

Introduction

1   Thomas Cowan, MD, "Covid-19/Coronavirus Caused By 5G?" https://www.brighteon.com/c32af45d-175c-4880-8398-938fb3483122.

Chapter 1

1   R. Koch, "Ueber den augenblicklichen Stand der bakteriologischen Choleradiagnose" [About the instantaneous state of the bacteriological diagnosis of cholera], *Zeitschrift für Hygiene und Infektionskrankheiten* (in German) 14 (1893): 319–38, doi:10.1007/BF02284324.
2   RJ Huebner, "Criteria for etiologic association of prevalent viruses with prevalent diseases; the virologist's dilemma," *Annals of the New York Academy of Sciences* 67, no. 8 (April 1957): 430-8. Bibcode:1957NYASA..67..430H. doi:10.1111/j.1749-6632.1957.tb46066.x. PMID 13411978; AS Evans, "Causation and disease: a chronological journey," (The Thomas Parran Lecture, 1978) *American Journal of Epidemiology* 142, no. 11 (December 1995): 1126–35, discussion1125. doi:10.1093/oxfordjournals.aje.a117571. PMID 7485059.
3   R. Fouchier et al. "Koch's postulates fulfilled for SARS virus," Nature 423, no. 240 (2003), discussed at https://www.youtube.com/watch?v=HsYjW0fNphA.

4   "Infectious Diseases at the Edward Worth Library," https://infectiousdiseases .edwardworthlibrary.ie/Theory-of-Contagion/.

5   Gerald L. Geison, *The Private Science of Louis Pasteur* (Princeton, NJ: Princeton University Press, 2014).

Chapter 2

1   John L. Heilbron, *Electricity in the 17th and 18th Centuries: A Study of Early Modern Physics* (Berkeley, CA: University of California Press, 1979), 490–491, quoted in Arthur Firstenberg, *The Invisible Rainbow: A History of Electricity and Life* (Santa Fe, NM: AGB Press, 2020), 32.

2   Frances Lowndes, *Observations on Medical Electricity* (London: D. Stuart, 1787), 39–40, quoted in Firstenberg, 32.

3   Heinrich Schweich, *Die Influenza: Ein historischerund atiologischer Versuch* (Berlin: Theodor Christian Friedrich Enslin), quoted in Firstenberg, 84.

4   Firstenberg. *The Invisible Rainbow*.

5   Ibid, 51-52.

6   Ibid, 85.

7   William Ian Beveridge, *Influenza: The Last Great Plague* (New York, NY: Prodist, 1978), 35, quoted Firstenberg, 86.

8   "1918 Flu Pandemic," https://www.history.com/topics/world-war-i/1918-flu-pandemic.

9   Ibid.

10   MJ Rosenau, "Experiments to Determine Mode of Spread of Influenza," *Journal of the American Medical Association* 73, no. 5 (August 2, 1919): 311–313.

11   Firstenberg, *The Invisible Rainbow*, 109.

12   Ibid, 111.

13   Thomas S. Cowan, MD, *Cancer and the New Biology of Water* (Hartford, VT: Chelsea Green, 2019)..

14   Firstenberg, *The Invisible Rainbow*, 369.

15   THP Nguyen et al, "The effect of a high frequency electromagnetic field in the microwave range on red blood cells," *Scientific Reports* 7, Article number: 10798 (2017), https://www.researchgate.net/publication/251830393_Cell_Effects_of _Electromagnetic_Radiation.

16   Shigeaki (Shey) Hakusui, "Wireless at 60 GHz Has Unique Oxygen Absorption Properties," *Scientists for Wired Technologies*, https://scientists4wiredtech.com /wireless-at-60-ghz-has-unique-oxygen-absorption-properties/.

17   "Central China province launches commercial 5G applications," XinHua Net, October 31, 2019, http://www.xinhuanet.com/english/2019-10/31/c_138517734.htm.

18   "THE 5G CORONAVIRUS SYDROME - All Mapped Out," March 17, 2020, https://weatherpeace.blogspot.com/2020/03/the-5g-coronavirus-sickness-mapped -out.html.

19   "San Marino 5G leader in Europe: first services launched" (May 23, 2018), https:// www.telecomitalia.com/en/press-archive/market/2018/PR-San-Marino5G-230518 .html.

20   Bartomeu Payeras I Cifre, "Study of the correlation between cases of coronavirus and the presence of 5G networks," trans. Claire Edwards (March-April 2020), www.tomeulamo.com/fitxers/264_CORONA-5G-d.pdf.

21   Ibid.

22 "Global Agendas Exposed" (March 17, 2020).
23 I Cifre, "Study of the correlation between cases of coronavirus and the presence of 5G networks."
24 Ibid.
25 Jeremy Kryt, "Will COVID-19 Wipe Out the Tribes of the Amazon?" *The Daily Beast*, May 25, 2020, https://www.thedailybeast.com/will-covid-19-wipe-out-the -tribes-of-the-amazon.
26 "Vivo Deploys 100G Network Across Amazon," *Light Reading*, October 11, 2013, https://www.lightreading.com/optical/100g/vivo-deploys-100g-network-across -amazon/d/d-id/706040.
27 Robert J. Burrowes, "Deadly rainbow: Will 5G precipitate the extinction of all life on Earth?" *Nation of Change*, July 7, 2020, https://www.nationofchange.org /2020/07/07/deadly-rainbow-will-5g-precipitate-the-extinction-of-all-life-on-earth/.
28 Pivotal Commware, Pivotal Echo 5G™, https://pivotalcommware.com/echo-5g/.
29 MedallionNet™ The Best Wi-Fi at Sea, https://www.princess.com/ships-and -experience/ocean-medallion/medallionnet/.
30 Staff Sgt. William Skelton, "New Marine Corps non-lethal weapon heats things up," *DVIDS*, March 9, 2012, https://www.dvidshub.net/news/85028/new-marine -corps-non-lethal-weapon-heats-things-up; Ross Kerber, "Ray gun, sci-fi staple, meets reality," *Boston Globe*, September 24, 2004.
31 I Belyaev et al, "EUROPAEM EMF Guideline 2016 for the Prevention, Diagnosis and Treatment of EMF-related Health Problems and Illnesses," *Rev Environ Health* 31, no. 3 (September 1, 2016): 363-97, doi: 10.1515/reveh-2016-0011.
32 RN Kostoff et al, "Adverse health effects of 5G mobile networking technology under real-life conditions," *Toxicology Letters* 323 (May 1, 2020): 35–40.
33 O. Johansson, "Disturbance of the Immune System by Electromagnetic Fields— A Potentially Underlying Cause for Cellular Damage and Tissue Repair Reduction Which Could Lead to Disease and Impairment," *Pathophysiology* 16, no. 2–3 (August 2009):157–77, doi: 10.1016/j.pathophys.2009.03.004. Epub 2009 Apr 23.
34 NP Zalyubovskaya and RI Kiselev, "Effect of Radio Waves of a Millimeter Frequency Range on the Body of Man and Animals," *Gigiyena I Sanitariya*, no. 8 (1978).
35 "Bibliography of Reported Biological Phenomena ('Effects') and Clinical Manifestations Attributed to Microwave and Radio-Frequency Radiation," Report, no. 1, MF12.524.015-0004B.
36 Dirk K F Meijer, Hans J. H. Geesink, and Jos Timmer, "The 5G Safety Dilemma: Plea for Urgent Scientific Research in the European Context" (April 2020), https:// www.researchgate.net/publication/340528995_The_5G_Safety_Dilemma_Plea _for_Urgent_Scientific_Research_in_the_European_Context.
37 L Giuranno L et al, "Radiation-Induced Lung Injury (RILI)," *Front Oncol* 9 (2019): 877. Published online 2019 Sep 6. doi: 10.3389/fonc.2019.00877.
38 "Lloyd's of London Insurance Won't Cover Smartphones – WiFi – Smart Meters – Cell Phone Towers By Excluding ALL Wireless Radiation Hazard," RF Safe, March 18, 2015, shttps://www.rfsafe.com/lloyds-of-london-insurance-wont -cover-smartphones-wifi-smart-meters-cell-phone-towers-by-excluding-all-wireless -radiation-hazards/.
39 "Dr. Cameron Kyle-Sidell Gives Thoughts On Current Global Pandemic," https://

search.aol.com/aol/video;_ylt=AwrE196Mue9e7V4AawtpCWVH;_
ylu=X3oDMTB0N2Noc21lBGNvbG8DYmYxBHBvcwMxBHZ0a
WQDBHNlYwNwaXZz?q=Sidell&s_it=searchtabs&v_t=lokiinbox#id=
1&vid=6e489b61db15c88fe6f7db58427fd840&action=view.

Chapter 3

[1]   Roberta JM Olson and Jay M Pasachoff, *Cosmos: The Art and Science of the Universe*
      (Islington, London: Reaktion Books, 2019).
[2]   Li Ch'un Feng, Director, Chinese Imperial Astronomical Bureau, 648 AD, quoted
      from https://www.researchgate.net/publication/326160954_Comets_and_
      Contagion_Evolution_Plague_and_Diseases_From_Space.
[3]   Rhawn Joseph PhD, Rudolf Schild PhD, & Chandra Wickramasinghe PhD,
      Biological Cosmology, Astrobiology, and the Origins and Evolution of Life (Cosmology
      Science Publishers, 2010), quoted in Gabriela Segura, MD, "New Light on the
      Black Death," *The Dot Connector Magazine* 13, no 1 (2011), https://health-matrix
      .net/2011/05/11/new-light-on-the-black-death-the-viral-and-cosmic-connection/.
[4]   Michelle Ziegler, "Procopius' Account of the Plague in Constantinople, 542," July
      31, 2011, https://hefenfelth.wordpress.com/2011/07/31/procopius-account-of-the
      -plague-in-constantinople-542/.
[5]   Dr. Marc Barton, "PLAGUES, COMETS AND VOLCANOES," Past Medical
      History, June 28, 2016, https://www.pastmedicalhistory.co.uk/plagues-comets-and
      -volcanoes/.
[6]   Michelle Ziegler, "Procopius' Account of the Plague in Constantinople, 542."
[7]   "Panspermia and the Origin of Life on Earth," https://www.panspermia-theory.com/.
[8]   Wal Thornhill, "Comets Impact Cosmology," July 20, 2004, https://www
      .holoscience.com/wp/comets-impact-cosmology/; Wallace Thornhill and David
      Talbott, "The Electric Comet," 2006, https://www.bibliotecapleyades.net/electric
      _universe/esp_electricuniverse17.htm.
[9]   Dr. Marc Barton, "PLAGUES, COMETS AND VOLCANOES."
[10] Thomas Short, *A general Chronological History of the Air, Weather, Seasons, Meteors*
      (London, 1749).
[11] Susan Scott & Christopher Duncan, *Return of the Black Death: The World's Greatest
      Serial Killer* (Wiley, 2004).
[12] Ibid.
[13] Mike Baillie, *New Light on the Black Death: The Cosmic Connection*, 1st ed. (History
      Press, 2006).
[14] David Meyer, "Did a Comet cause the Black Death?" July 11, 2011, www
      .davidmeyercreations.com/strange-science/did-a-comet-cause-the-black-death/.
[15] S Likitvivatanavong et al, "Multiple receptors as targets of Cry toxins in
      mosquitoes," *J Agric Food Chem* 59, no. 7 (April 13, 2011): 2829–2838. Published
      online 2011 Jan 6. doi: 10.1021/jf1036189.
[16] H Batliwala et al, "Methane-induced hemolysis of human erythrocytes,"
      Biochemical Journal 307 (1995): 433–438..
[17] https://www.amazon.com/Ultrasound-Causation-Microcephaly-Virus-Hypothesis
      /dp/1941719082..
[18] Eleanor Herman, *The Royal Art of Poison: Filthy Palaces, Fatal Cosmetics, Deadly
      Medicine, and Murder Most Foul*, 1st ed. (St. Martin's Press, June 12, 2018).

[19] Tamara Bhandari, "Why people with type O blood are more likely to die of cholera," https://www.sciencedaily.com/releases/2016/08/160829105908.htm.

[20] Judith Summers, *Soho—A History of London's Most Colourful Neighborhood* (London: Bloomsbury, 1989)..

[21] Ibid, 113–117.

[22] TJ Inglis, "Principia aetiologica: taking causality beyond Koch's postulates," *Journal of Medical Microbiology* 56, Pt 11 (November 2007): 1419–22. doi:10.1099/jmm .0.47179-0. PMID 17965339.

[23] Ron Schmid, ND, "PASTEURIZE OR CERTIFY: TWO SOLUTIONS TO 'THE MILK PROBLEM,'" A Campaign for Real Milk, December 13, 2003, https ://www.realmilk.com/safety/pasteurize-or-certify/.

[24] https://en.wikipedia.org/wiki/Smallpox

[25] Charles AR Campbell, MD, *Resume of Experiments on Variola* (San Antonio, Texas), http://whale.to/a/campbell1.html.

[26] Ibid.

[27] Ibid.

[28] Kaushik Patowary, "Dr. Charles Campbell And His Malaria-Fighting Bat Towers," *Amusing Planet*, https://www.amusingplanet.com/2018/09/dr-charles-campbell -and-his-malaria.html.

[29] TJ Inglis, "Principia aetiologica: taking causality beyond Koch's postulates."

[30] G Bordenave, "Louis Pasteur (1822-1895)," *Microbes and Infection* 5, no. 6 (May 2003): 553–60. doi:10.1016/S1286-4579(03)00075-3. PMID 12758285.

[31] A. Sakula, "Robert Koch: centenary of the discovery of the tubercle bacillus, 1882," *Thorax* 37, no. 4 (April 1982): 246–251. doi: 10.1136/thx.37.4.246.

[32] Robert Koch, "The Etiology of Tuberculosis," *Reviews of Infectious Diseases* 4, no. 6 (November 1982): 1270–1274, https://doi.org/10.1093/clinids/4.6.1270.

[33] Weston A. Price, *Nutrition and Physical Degeneration* (Price-Pottenger Nutrition Foundation, 1945).

[34] Ibid, 331.

[35] Ibid, 42.

[36] Ibid,. 51.

[37] Ibid, 130–133.

[38] "Toxicological Profile: for DDT, DDE, and DDE," Agency for Toxic Substances and Disease Registry, September 2002; "DDT. Immediately Dangerous to Life and Health Concentrations (IDLH)," National Institute for Occupational Safety and Health (NIOSH).

[39] NobelPrize.org: The Nobel Prize in Physiology of Medicine 1948.

[40] Jim West, "Pesticides and Polio: A Critique of Scientific Literature," The Weston A. Price Foundation, February 8, 2003, https://www.westonaprice.org/health-topics /environmental-toxins/pesticides-and-polio-a-critique-of-scientific-literature/

[41] Compiled by Jim West from *US Vital Statistics*, US Government Printing Office, Washington, DC; published in "Pesticides and Polio."

[42] Jim West, "Pesticides and Polio: A Critique of Scientific Literature."

[43] Torsten Engelbrecht and Claus Kohnlein, Virus Mania (Trafford Publishing, 2007) 66.

[44] Agnes Ullmann, "Pasteur–Koch: Distinctive Ways of Thinking about Infectious Diseases," *Microbe* 2, no. 8 (2007): 383–7. Archived from the original on 2011-07-22.

[45] Dawn Lester and David Parker, *What Really Makes You Ill? Why Everything You Thought You Knew About Disease Is Wrong* (Independently published, 2019).

[46] WG Winkler, "Airborne Rabies Virus Isolation," *Bulletin of the Wildlife Disease Association* 4, no. 2 (December 12, 1967): 37-40.

[47] DM Pastula et al, "Acute neurologic illness of unknown etiology in children - Colorado, August-September 2014 (PDF)," *MMWR Morb. Mortal.* Wkly. Rep. 63, no. 40 (October 10, 2014): 901–2. PMC 4584613. PMID 25299607.

[48] R Dhiman et al, "Correlation between non-polio acute flaccid paralysis rates with pulse polio frequency in India," *Int J Environ Res Public Health* 15, no. 8 (2018).

[49] S Humphries and R Bystrianyk, "The 'disappearance' of polio," *Dissolving Illusions: Disease, Vaccines, and the Forgotten History* (Independently published, 2013) 222-92.

[50] Xcvi Raymond Obomsawin, "Historical and scientific perspectives on the health of Canada's First peoples," https://www.worldcat.org/title/historical-and-scientific -perspectives-on-the-health-of-canadas-first-peoples/oclc/855308523; https://www .soilandhealth.org/wp-content/uploads/02/0203CAT/020335.obomsawin.pdf.

[51] Ibid, 14.

[52] National Commission Inquiry on Indian Health, *The History of Indian Health*, 6–7.

[53] AR Bharti et al, "Leptospirosis: a zoonotic disease of global importance," *The Lancet Infectious Diseases* 3, no. 12 (December 1, 2003): 757–771, https://www.thelancet .com/journals/laninf/article/PIIS1473-3099(03)00830-2/fulltext.

[54] Garcilaso de la Vega, *The Florida of the Inca* (Austin, TX: University of Texas Press, 1951) 421.

[55] PM Kraemer, "New Mexico's Ancient Salt Trade," http://www.elpalacio.org/articles /winter12/salttrade-v82-no1.pdf. Accessed June 21, 2017.

[56] Royal Commission on Aboriginal Peoples (Canada), 1996, http://data2.archives .ca/e/e448/e011188230-01.pdf.

[57] Dave Mihalovic, "Biologist wins Supreme Court case proving that the measles virus does not exist," *Signs of the Times*, January 27, 2017, https://www.sott.net /article/340948-Biologist-wins-Supreme-Court-case-proving-that-the-measles -virus-does-not-exist.

[58] James Herer, "Microbiologist and Virologist Dr. Stefan Lanka: 'Viruses Do Not Cause Diseases and Vaccines are Not Effective,'" *Weblyf*, https://www.weblyf.com /2020/05/microbiologist-and-virologist-dr-stefan-lanka-viruses-do-not-cause- diseases-and-vaccines-are-not-effective/.

[59] "Anti-Vaxxer Biologist Stefan Lanka Bets Over $100K Measles Isn't A Virus; He Wins In German Federal Supreme Court," January 21, 2017, https://anonhq.com /anti-vaxxer-biologist-stefan-lanka-bets-100k-measles-isnt-virus-wins-german -federal-supreme-court/.

Chapter 4

[1] Torsten Engelbrecht and Claus Kohnlein, Virus Mania, 21

[2] Ibid, 90.

[3] PM Sharp and BH Hahn, "Origins of HIV and the AIDS pandemic," *Cold Spring Harbor Perspectives in Medicine* 1, no. 1 (September 2011), a006841. doi:10.1101 /cshperspect.a006841. PMC 3234451. PMID 22229120.

[4] The three best are books on this subject are *Virus Mania* by Torsten Engelbrecht

and Claus Kohnlein; *The Silent Revolution in AIDS and Cancer* by Heinrich Kremmer; and *AIDS, Opium, Diamonds and Empire* by Nancy Banks.

5   NS Padian et al, "Heterosexual Transmission of Human Immunodeficiency Virus (HIV) in Northern California: Results From a Ten-Year Study," *Am J Epidemiol* 146, no. 4 (August 15, 1997): 350-7. doi: 10.1093/oxfordjournals.aje.a009276.

6   M Fioranelli et al, "5G Technology and induction of coronavirus in skin cells," *Journal of Biological Regulators & Homeostatic Agents* 54, no. 4 (June 9, 2020).

7   Simon Garfield, "The rise and fall of AZT: It was the drug that had to work. It brought hope to people with HIV and AIDS, and millions for the company that developed it. It had to work. There was nothing else. But for many who used AZT - it didn't," *The Independent*, May 2, 1993, https://www.independent.co.uk/arts-entertainment/the -rise-and-fall-of-azt-it-was-the-drug-that-had-to-work-it-brought-hope-to-people -with-hiv-and-2320491.html.l

8   Torsten Engelbrecht and Claus Kohnlein, *Virus Mania*, 11.

9   Peng Zhou et al, "Discovery of a novel coronavirus associated with the recent pneumonia outbreak in humans and its potential bat origin," *bioRxiv*. doi: https:// doi.org/10.1101/2020.01.22.914952; Na Zhu et al, "A Novel Coronavirus from Patients with Pneumonia in China, 2019," *N Engl J Med* 382 (February 20, 2020) 727-733, DOI: 10.1056/NEJMoa2001017.; Jeong-Min Kim et al, "Identification of Coronavirus Isolated From a Patient in Korea With COVID-19," *Osong Public Health Res Perspect*.11, no. 1 (February 2020): 3-7.doi: 10.24171/j. phrp.2020.11.1.02.; Karen Mossman, "I study viruses: How our team isolated the new coronavirus to fight the global pandemic," McMaster University, March 25, 2020, https://brighterworld.mcmaster.ca/articles/i-study-viruses-how-our-team -isolated-the-new-coronavirus-to-fight-the-global-pandemic/.

10 "The Rooster in the River of Rats," by Andrew Kaufman, MD, https://www .youtube.com/watch?v=NTws_mAsDfU,.

11 R AM Fouchier et al, "Koch's postulates fulfilled for SARS virus," *Nature* 423 (2003): 240.

12 JFW Chan et al, "Simulation of the Clinical and Pathological Manifestations of Coronavirus Disease 2019 (COVID-19) in Golden Syrian Hamster Model: Implications for Disease Pathogenesis and Transmissibility," *Clin Infect Dis.* (March 26, 2020), ciaa325. doi: 10.1093/cid/ciaa325.

Chapter 5

1   C. Huang et al, "Clinical features of patients infected with 2019 novel coronavirus in Wuhan, China," *The Lancet* (January 24, 2020), https://www.thelancet.com /journals/lancet/article/PIIS0140-6736(20)30183-5/fulltext 3.

2   David Crowe, *Flaws in Coronavirus Pandemic Theory*, 5, https://theinfectiousmyth .com/book/CoronavirusPanic.pdf.

3   Peter Fimrite, "Studies show coronavirus antibodies may fade fast, raising questions about vaccines," *San Francisco Chronicle*, July 17, 2020, https://www.sfchronicle .com/health/article/With-coronavirus-antibodies-fading-fast-focus-15414533.php.

4   "Clinical Questions about COVID-19: Questions and Answers," https://www.cdc .gov/coronavirus/2019-ncov/hcp/faq.html. Accessed July 26, 2020.

5   Erika Edwards, Courtney Kube and Mark Schone, "Some have tested positive for

COVID-19 after recovering. What does that mean?" NBC News, May 19, 2020, https://news.yahoo.com/tested-positive-covid-19-recovering-223200125.html.

6   Ibid.

7   http://agenda-leben.de/Lanka_Diplomarbeit_1989_kompr.pdf.

8   Torsten Engelbrecht and Konstantin Demeter, "COVID-19 PCR Tests Are Scientifically Meaningless," *OffGuardian*, June 27, 2020, https://off-guardian.org/2020/06/27/covid19-pcr-tests-are-scientifically-meaningless/?fbclid=IwAR0OFMLQ-oW85YSrDczm8rjLC1cCJmJ4lIIoW3_-PIYYJRypsmgh2CH8fJ4.

9   L Leo et al, "Emergence of a novel human Coronavirus threatening human health," *Nature Medicine* (March 2020)..

10 Myung-guk Han et al, "Identification of Coronavirus Isolated from a Patient in Korea with COVID-19," *Song Public Health and Research Perspectives* (February 2020).

11 Wan Beom Park et al, "Virus Isolated from the First Patient with SARS-CoV-2 in Korea," *Journal of Korean Medical Science* (February 24, 2020).

12 Na Zhu et al, "A Novel Coronavirus from Patients with Pneumonia in China, 2019," *New England Journal of Medicine* (February 20, 2020).

13 Danielle Wallace, "Ventura County clarifies claims it would force people from homes into isolated coronavirus centers," Fox News, May 7, 2020, https://www.foxnews.com/us/california-ventura-county-coronavirus-forcibly-removed-homes-quarantine.

14 Adrianna Rodriguez, "'Heartbreaking': Moms could be separated from their newborns under coronavirus guidelines," *USA Today*, March 26, 2020, https://www.usatoday.com/story/news/health/2020/03/26/pregnant-women-covid-19-could-separated-babies-birth/2907751001/.

15 Jessica Lee, "Did Tanzania's President Expose Faulty COVID-19 Testing by Submitting Non-Human Samples?" *Snopes*, May 7, 2020, https://www.snopes.com/fact-check/tanzania-president-covid-tests/.

16 James Herer, "Coronavirus: The Truth about PCR Test Kit from the Inventor and Other Experts," *Weblyf*, https://www.weblyf.com/2020/05/coronavirus-the-truth-about-pcr-test-kit-from-the-inventor-and-other-experts/.

17 "CDC 2019-Novel Coronavirus (2019-nCoV) Real-Time RT-PCR Diagnostic Panel," Centers for Disease Control and Prevention, https://www.fda.gov/media/134922/download.

18 "Accelerated Emergency Use Authorization (Eua) Summary Covid-19 Rt-Pcr Test (Laboratory Corporation Of America)," https://www.fda.gov/media/136151/download

19 http://technical-support.roche.com/_layouts/net.pid/Download.aspx?documentID=1cca7ff9-388a-ea11-fa90-005056a772fd&fileName=TP00886v2&extension=pdf&mimeType=application%2Fpdf&inline=False

20 David Crowe, "Antibody Testing for COVID-19," May 13, 2020, https://theinfectiousmyth.com/coronavirus/AntibodyTestingForCOVID.pdf.

21 F Zhou et al, "Clinical course and risk factors for mortality of adult inpatients with COVID-19 in Wuhan, China: a retrospective cohort study," *The Lancet* (March 11, 2020), https://www.thelancet.com/journals/article/PIIS0140-6736(20)30566-3/fulltext. .

22 R. Prasad, "Meta-analysis does not support continued use of point-of-care serological tests for COVID-19," The Hindu, July 4, 2020, https://www.thehindu

.com/sci-tech/science/meta-analysis-does-not-support-continued-use-of-point-of
-care-serological-tests-for-covid-19/article31989748.ece.

Chapter 6

1   G. Bordenave, "Louis Pasteur (1822–1895)," *Microbes and Infection / Institut Pasteur* 5, no. 6 (May 2003): 553–60, doi:10.1016/S1286-4579(03)00075-3.

2   "Dr. Stefan Lanka Debunks Pictures of 'Isolated Viruses,'" Vaccination Information Network, https://www.vaccinationinformationnetwork.com/dr-stefan-lanka-debunks -pictures-of-isolated-viruses/.

3   MD Keller et al, "Decoy exosomes provide protection against bacterial toxins," *Nature* 579 (2020): 260–264 (2020); "Newfound Cell Defense System Features Toxin-Isolating 'Sponges,'" Yahoo Finance, March 4, 2020, https://finance.yahoo .com/news/newfound-cell-defense-system-features-160000044.html.

4   G Pironti et al, "Circulating Exosomes Induced by Cardiac Pressure Overload Contain Functional Angiotensin II Type 1 Receptors," *Circulation* no. 131 (2015): 2120–2130, Originally published 20 May 2015, https://doi.org/10.1161/ CIRCULATIONAHA.115.015687..

5   William A. Wells, "When is a virus an exosome?" *Journal of Cell Biology* 162, no. 6 (2003): 960, https://rupress.org/jcb/article/162/6/960/33690/When-is-a-virus-an -exosome.

6   "Newfound Cell Defense System Features Toxin-Isolating 'Sponges,'" Yahoo Finance, March 4, 2020, https://finance.yahoo.com/news/newfound-cell-defense -system-features-160000044.html.

7   G Raposo and W Stoorvogel, "Extracellular vesicles: Exosomes, microvesicles, and friends," *J Cell Biol* 200, no. 4 (February 18, 2013): 373–383, doi: 10.1083 /jcb.201211138.

8   C Frühbeis et al, "Extracellular vesicles as mediators of neuron-glia communication," *Front Cell Neurosci* 7 (2013): 182. Published online 2013 Oct 30. doi: 10.3389 /fncel.2013.00182.

9   OD Mrowczynski et al, "Exosomes impact survival to radiation exposure in cell line models of nervous system cancer," *Oncotarget* 9, no. 90 (November 16, 2018): 36083–36101. Published online 2018 Nov 16, doi: 10.18632/oncotarget.26300.

10   https://newumedspaorlando.com/exosomes-penis-treatment-orlando/.

11   J Smythies J et al, "Molecular mechanisms for the inheritance of acquired characteristics—exosomes, microRNA shuttling, fear and stress: Lamarck resurrected?" *Front Genet* 5 (2014): 133. Published online 2014 May 15. Prepublished online April 16, 2014. doi: 10.3389/fgene.2014.00133; KeFang et al, "Differential serum exosome microRNA profile in a stress-induced depression rat model," *Journal of Affective Disorders* 274 (September 1, 2020):144–158, https:// doi.org/10.1016/j.jad.2020.05.017.

12   YE Young-Eun Cho et al, "Exosomes: An emerging factor in stress-induced immunomodulation," *Seminars in Immunology* 26, no. 5 (October 2014): 394–401.

13   W Seo et al, "Exogenous exosomes from mice with acetaminophen-induced liver injury promote toxicity in the recipient hepatocytes and mice," *Scientific Reports* 8, Article number: 16070 (2018), Published: October 30, 2018.

Chapter 7

1   Dan Evon, "Did This Nobel Prize Winner Say COVID-19 Was Created in a Lab?" *Snopes*, April 29, 2020, https://www.snopes.com/fact-check/luc-montagnier-covid-created-lab/.
2   L Montagnier et al, "Electromagnetic Signals Are Produced by Aqueous Nanostructures Derived From Bacterial DNA Sequences," *Interdiscip Sci.* 1, no. 2 (June 2009): 81-90. doi: 10.1007/s12539-009-0036-7. Epub March 4, 2009.
3   "Childhood Infectious Diseases Protect Us From Cancer Later In Life," http://vaxinfostarthere.com/childhood-infectious-diseases-protect-us-cancers-later-life/.

Chapter 8

1   Gerald Pollack, *Cells, Gels and the Engines of Life* (Seattle, WA: Ebner & Sons, 2001).
2   https://cassiopaea.org/forum/threads/gerald-pollack-electrically-structured-water.31363/page-4
3   H Yoo et al, "Contraction-Induced Changes in Hydrogen Bonding of Muscle Hydration Water," *J Phys Chem Lett.* 5, no. 6 (March 20, 2014): 947–952. Published online 2014 Feb 25. doi: 10.1021/jz5000879.
4   Personal communication with Gerald Pollack, PhD, July 7, 2020.
5   "Father Richard Willhelm H2O2 Lourdes Water has extra Oxygen not extra Hydrogen," https://www.youtube.com/watch?v=8F3sTiBC6uY.
6   Stacey A. Reading and Maggie Yeomans, "Oxygen absorption by skin exposed to oxygen supersaturated water," *Canadian Journal of Physiology and Pharmacology* 90, no. 5 (2012): 515-524, https://doi.org/10.1139/y2012-020.
7   ST Kyoren et al, "Effect of high consentrated dissolved oxygen on the plant growth in a deep hydroponic culture under a low temperature," *IFAC Proceedings* 43, no. 26 (2010): 251–255; "Dissolved Oxygen for Better Growth: Part I: What Is It and Why Do Plants Need It?" https://www.questclimate.com/dissolved-oxygen-better-growth-part-plants-need/.
8   Daniel Ladizinsky, MD and David Roe, PhD, "New Insights Into Oxygen Therapy for Wound Healing," *Wounds* 22, no. 12 (2010): 294300..
9   N Fleming et al, "Ingestion of oxygenated water enhances lactate clearance kinetics in trained runners," *Journal of the International Society of Sports Nutrition* 14, no. 9 (2017), DOI 10.1186/s12970-017-0166-y.
10 R Grubera et al, "The influence of oxygenated water on the immune status, liver enzymes, and the generation of oxygen radicals: a prospective, randomized, blinded clinical study," *Clin Nutr.* 24, no. 3 (June 2005): 407–14, https://doi.org/10.1016/j.clnu.2004.12.007.
11 MV Ivannikov et al, "Neuromuscular Transmission and Muscle Fatigue Changes by Nanostructured Oxygen," *Muscle and Nerve* 55, no. 4 (April 2017): 555–563. PMID: 27422738 • DOI: 10.1002/mus.25248.
12 "Cancer metastasis: The unexpected perils of hypoxia," *Science News*, Ludwig-Maximilians-Universität München, May 22, 2017.

Chapter 9

[1] P Le Pogam et al, "Untargeted metabolomics unveil alterations of biomembranes permeability in human HaCaT keratinocytes upon 60 GHz millimeter-wave exposure," *Sci Rep.* 9 (2019): 9343. doi: 10.1038/s41598-019-45662-6; PH Siegel and V Pikov, "Impact of Low Intensity Millimeter-Waves on Cell Membrane Permeability," October 2009, DOI: 10.1109/ICIMW.2009.5325755. *IEEE Xplore.*

[2] Sally Fallon Morell, *Nourishing Fats.* Grand Central, New York, 2017, pp. 85–86.

[3] Ibid.

[4] Ibid, 113-114.

[5] Joaquin Timoneda et al, "Vitamin A Deficiency and the Lung," *Nutrients* 10, no. 9 (August 21, 2018): 1132. doi: 10.3390/nu10091132.

[6] Sally Fallon and Mary G. Enig, PhD, "Be Kind to Your Grains ... And Your Grains Will Be Kind To You," The Weston A. Price Foundation, January 1, 2000, https://www.westonaprice.org/health-topics/food-features/be-kind-to-your-grains-and-your-grains-will-be-kind-to-you/.

[7] George Washington, *To Make Small Beer.* George Washington Papers, 1757, New York Public Library Archive.

[8] Weston A. Price Foundation, "Dirty Secrets of the Food Processing Industry," December 26, 2005, https://www.westonaprice.org/health-topics/modern-foods/dirty-secrets-of-the-food-processing-industry/.

[9] L Blandón-Naranjo et al, "Electrochemical Behaviour of Microwave-assisted Oxidized MWCNTs Based Disposable Electrodes: Proposal of a NADH Electrochemical Sensor," *Electroanalysis* (January 16, 2018).

[10] F Ameer et al, "De novo lipogenesis in health and disease," *Metabolism* 63, no. 7 (July 2014): 895–902.

[11] http://es-forum.com/How-I-Healed-EMF-Sensitivity-td4030455.html.

[12] Sally Fallon Morell, "New Evidence That Processing Destroys Milk Proteins," March 21, 2020, https://www.realmilk.com/health/new-evidence-that-processing-destroys-milk-proteins/; https://www.realmilk.com/wp-content/uploads/2020/06/CampaignforRealMilkSept2011PPTasPDF.pdf, 4–12.

[13] Sally Fallon Morell, "What Pasteurization Does To The Vitamins In Milk," October 31, 2018, https://www.realmilk.com/health/pasteurization-vitamins-milk.

[14] HM Said et al, "Intestinal Uptake of Retinol: Enhancement by Bovine Milk Beta-Lactoglobulin," *Am J Clin Nutr.* 49, no. 4 (April 1989): 690–4. doi: 10.1093/ajcn/49.4.690..

[15] B Sozańska, "Raw Cow's Milk and Its Protective Effect on Allergies and Asthma," *Nutrients* 11, no. 2 (February 2019): 469. Published online Feb. 22, 2019. doi: 10.3390/nu11020469; https://www.realmilk.com/health/raw-milk-protective-against-asthma-and-allergies/.

[16] Alexey V Polonikov, "Endogenous deficiency of glutathione as the most likely cause of serious manifestations and death in patients with the novel coronavirus infection (COVID-19): a hypothesis based on literature data and own observations," https://www.researchgate.net/publication/340917045_Endogenous_deficiency_of_glutathione_as_the_most_likely_cause_of_serious_manifestations_and_death_in_patients_with_the_novel_coronavirus_infection_COVID-19_a_hypothesis_based_on_literature_data_and_ow.

17 Sally Robertson, BSc, "Study links fermented vegetable consumption to low
    COVID-19 mortality," *News-Medical*.net, July 8, 2020, https://www.news-medical
    .net/news/20200708/Study-links-fermented-vegetable-consumption-to-low-COVID
    -19-mortality.aspx..

18 "Ilya Mechnikov – Biographical," *Nobelprize.org*. Nobel Media AB.

19 Louisa Williams, ND, "Dr. Ilya Metchnikoff Drank Cholera and Lived!" April 17,
    2020, https://www.louisawilliamsnd.com/post/dr-ilya-metchnikoff.

20 Merinda Teller, MPh, PhD, "Debunking the Myth That Microwave Ovens Are
    Harmless," The Weston A. Price Foundation, November 5, 2019, https://www
    .westonaprice.org/health-topics/debunking-the-myth-that-microwave-ovens-are
    -harmless/..

Chapter 10

1   "Arsenic in drinking water seen as threat," *USAToday.com*, August 30, 2007;
    P Ravenscroft, "Predicting the global distribution of arsenic pollution in
    groundwater," Paper presented at: Arsenic – The Geography of a Global Problem,
    Royal Geographic Society Arsenic Conference held at: Royal Geographic Society,
    London, England, August 29, 2007.

2   FT Jones, "A Broad View of Arsenic," *Poult Sci*. 86, no. 1 (January 2007): 2–14.
    doi: 10.1093/ps/86.1.2.

3   "What's in your mouth....Mercury Fillings Smoking Teeth," https://www.youtube
    .com/watch?v=o2VCen1vCMY..

4   FDA, "Thimerosal in Vaccines," Archived from the original on October 26, 2006.

5   "FAQ's About Mercury (Thimerosal) in Vaccines," National Vaccine Information
    Center, https://www.nvic.org/faqs/mercury-thimerosal.aspx.

6   Jose Biller, *Interface of neurology and internal medicine* (Philadelphia, PA: Lippincott
    Williams & Wilkins, 2008) Chapter 163, 939.

7   S Mahernia et al, "Determination of hydrogen cyanide concentration in
    mainstream smoke of tobacco products by polarography," *J Environ Health Sci Eng*.
    13, no. 57 (2015). Published online July 29, 2015. doi: 10.1186/s40201-015
    -0211-1..

8   8 X Wu et al, "Exposure to air pollution and COVID-19 mortality in the United
    States: a nationwide cross-sectional study," medRxiv preprint. doi: https://doi.org
    /10.1101/2020.04.05.20054502.

9   O Johansson, "Disturbance of the Immune System by Electromagnetic fields—A
    Potentially Underlying Cause for Cellular Damage and Tissue Repair Reduction
    Which Could Lead to Disease and Impairment," *Pathophysiology* 16, no. 2-3 (August
    2009): 157-77. doi: 10.1016/j.pathophys.2009.03.004. Epub April 23, 2009.

10 LA Pushnoy et al, "Herbicide (Roundup) pneumonitis," *Chest* 114, no. 6 (1998):
    1769-71.

11 Stephanie Seneff, PhD, "Air Pollution, Biodiesel, Glyphosate and Covid-19," *Wise
    Traditions in Food, Farming and the Healing Arts* 21, no. 2 (Summer 2020).

12 J Gabbatiss, "Air pollution from UK shipping is four times higher than previously
    thought," *Independent*, February 3, 2018.

13 Sadiq Kahn, "Biodiesel and London buses," July 18, 2017, https://www.london.gov
    .uk/questions/2017/2662.

[14] M Lin and E Kao, "CPC to phase out B2 biodiesel in three months," *Focus Taiwan*, May 5, 2014.

[15] B Berke, "Interactive: an updated look at who coronavirus hits hardest in Massachusetts," *The Enterprise*, April 14, 2020.

[16] NL Swanson NL et al, "Genetically engineered crops, glyphosate and the deterioration of health in the United States of America," *J Org Syst* 9 (2014): 6-37.

[17] Stephanie Seneff, PhD, "Air Pollution, Biodiesel, Glyphosate and Covid-19," *Wise Traditions in Food, Farming and the Healing Arts* 21, no. 2 (Summer 2020): 29.

[18] Xingzhong Hu, Dong Chen, et al, "Low Serum Cholesterol Level Among Patients with COVID-19 Infection in Wenzhou, China," preprint with *The Lancet* (March 2, 2020), https://papers.ssrn.com/sol3/papers.cfm?abstract_id=3544826.

[19] "Nearly 7 in 10 Americans are on prescription drugs," Mayo Clinic, June 19, 2003, https://www.sciencedaily.com/releases/2013/06/130619132352.htm.

[20] Joe Graedon, "Lisinopril Side Effects Can Be Lethal," *The People's Pharmacy*, February 15, 2018, https://www.peoplespharmacy.com/articles/lisinopril-side -effects-can-be-lethal.

[21] James Franklin Lee Jr., "Aluminum, Barium, and Chemtrails Explained – JUST THE FACTS," Climate Viewer News, March 15 2015, https://climateviewer. com/2015/03/15/aluminum-barium-and-chemtrails-explained-just-the-facts/.

[22] Committee on Nutrition, "Soy Protein-based Formulas: Recommendations for Use in Infant Feeding," *Pediatrics* 101, no. 1 (January 1998): 148-153, DOI: https:// doi.org/10.1542/peds.101.1.148.

[23] C Exley et al, "Aluminum in Tobacco and Cannabis and Smoke Related Disease," *American Journal of Medicine* 119 (2006): 276.e9-276.e11.

[24] C Exley and E Clarkson, "Aluminium in human brain tissue from donors without neurodegenerative disease: A comparison with Alzheimer's disease, multiple sclerosis and autism," *Scientific Reports* 10, Article number: 7770 (2020).

[25] "Vaccine excipient and media summary. Excipients included in U.S. vaccines, by vaccine," Centers for Disease Control, https://www.cdc.gov/vaccines/pubs /pinkbook/downloads/appendices/b/excipient-table-2.pdf.

[26] Christina England, "The FDA Approves a New HPV Vaccine Containing Over Twice as Much Aluminum As its Predecessor," *VacTruth.com*, February 1, 2015, https://vactruth.com/2015/02/01/vaccine-containing-aluminum/.

[27] O Vera-Lastra et al, "Autoimmune/inflammatory Syndrome Induced by Adjuvants (Shoenfeld's Syndrome): Clinical and Immunological Spectrum," *Expert Rev Clin Immunol*, DOI: 10.1586/eci.13.2.

[28] Christopher Exley, PhD, FRSB, "Surviving in the Aluminum Age," The Weston A. Price Foundation, April 24, 2019, https://www.westonaprice.org/health-topics /environmental-toxins/surviving-in-the-aluminum-age/.

[29] Kendall Nelson, "Aluminum in Vaccines: What Everyone Needs to Know," The Weston A. Price Foundation, May 7, 2018, https://www.westonaprice.org/health -topics/vaccinations/aluminum-in-vaccines-what-everyone-needs-to-know/.

[30] G Wolff, "Influenza vaccination and respiratory virus interference among Department of Defense personnel during the 2017–2018 influenza season," Vaccine 38, no. 2 (January 10, 2020): 350–354.

[31] "Coverage rate of flu vaccination in Italy 2006-2019," Statista Research Department, March 23, 2020, https://www.statista.com/statistics/829799/coverage-rate-of-flu

-vaccination-in-italy/; C de Waure et al, "Adjuvanted influenza vaccine for the Italian elderly in the 2018/19 season: an updated health technology assessment," *European Journal of Public Health* 29, no. 5 (October 2019): 900–905, https://doi .org/10.1093/eurpub/ckz041.

32 "China: Vaccine Law Passed," Library of Congress Law, August 27, 2019, https:// www.loc.gov/law/foreign-news/article/china-vaccine-law-passed/?fbclid=IwAR35hj W8ev1pKHCtw138-w84y15TW2kX5P-8houXmAFaayUnZ_YPpYsmPU.

Chapter 12

1 "COVID-19 Dashboard by the Center for Systems Science and Engineering (CSSE) at Johns Hopkins University (JHU)," https://gisanddata.maps.arcgis.com /apps/opsdashboard/index.html#/bda7594740fd40299423467b48e9ecf6.

2 "Cases in the U.S.," Centers for Disease Control, Accessed July 30, 2020, https:// www.cdc.gov/coronavirus/2019-ncov/cases-updates/cases-in-us.html?CDC_AA_ref Val=https%3A%2F%2Fwww.cdc.gov%2Fcoronavirus%2F2019-ncov%2Fcases-updates%2Fsummary.html.

3 "Common Human Coronaviruses," Centers for Disease Control, https://www.cdc .gov/coronavirus/general-information.html.

4 "More Than 40% of U.S. Coronavirus Deaths Are Linked to Nursing Homes," *New York Times*, Updated July 23, 2020, nytimes.com/interactive/2020/us /coronavirus-nursing-homes.html.

5 Jim Hoft, "Is the Coronavirus or the 2019-2020 Flu More Dangerous for US Seniors? — Here are the Numbers," *Gateway Pundit*, March 16, 2020, https:// www.thegatewaypundit.com/2020/03/is-the-coronavirus-or-the-2019-2020-flu -more-dangerous-for-us-seniors-here-are-the-numbers/; Tommaso Ebhardt, Chiara Remondini, and Marco Bertacche, "99% of Those Who Died From Virus Had Other Illness, Italy Says," *Bloomberg*, March 18, 2020, https://www.bloomberg.com /news/articles/2020-03-18/99-of-those-who-died-from-virus-had-other-illness-italy -says.

6 James Barrett, "Stanford Professor: Data Indicates We're Severely Overreacting To Coronavirus," *Daily Wire*, March 18, 2020, https://www.dailywire.com/news /stanford-professor-data-indicates-were-overreacting-to-coronavirus.

7 Robert Preidt, "Study: Most N.Y. COVID Patients on Ventilators Died," Web MD, April 22, 2020, https://www.webmd.com/lung/news/20200422/most-covid-19 -patients-placed-on-ventilators-died-new-york-study-shows#1.

8 Jon Miltimore, "Physicians Say Hospitals Are Pressuring ER Docs to List COVID-19 on Death Certificates. Here's Why," Foundation for Economic Education, April 29, 2020, https://fee.org/articles/physicians-say-hospitals-are -pressuring-er-docs-to-list-covid-19-on-death-certificates-here-s-why/.

9 Ben Warren, "Official Raises Alarm on Inflated COVID-19 Deaths," *News Wars*, April 8, 2020, https://www.newswars.com/official-raises-alarm-on-inflated-covid -19-deaths/

10 https://healthfeedback.org/claimreview/mortality-in-the-u-s-noticeably-increased-during-the-first-months-of-2020-compared-to-previous-years/

11 Tedd Koren, D.C., "The nursing home pandemic," *Koren Wellness*, May 29, 2020, https://korenwellness.com/blog/iatrogenic-illness/.

12   Wang et al, "Remdesivir in adults with severe COVID-19: a randomised, double-blind, placebo-controlled, multicentre trial," *The Lancet* 395, no. 10236 (May 2020): 1569–1578, doi:10.1016/S0140-6736(20)31022-9. PMC 7190303. PMID 32423584.

13   S Richardson et al, "Presenting Characteristics, Comorbidities, and Outcomes Among 5700 Patients Hospitalized With COVID-19 in the New York City Area," *JAMA* 323, no. 20 (2020): 2052-2059. doi:10.1001/jama.2020.6775; Ariana Eunjung Cha, "In New York's largest hospital system, many coronavirus patients on ventilators didn't make it," Washington Post, April 26, 2020, https://www .washingtonpost.com/health/2020/04/22/coronavirus-ventilators-survival/.

14   S Richardson et al, "Presenting Characteristics, Comorbidities, and Outcomes Among 5700 Patients Hospitalized With COVID-19 in the New York City Area," *JAMA* 323, no. 20 (April 22, 2020): 2052-2059.

15   Ralph Ellis and Andrea Kane, "Pathologist found blood clots in 'almost every organ' during autopsies on Covid-19 patients," CNN, July 10, 2020, https://www.cnn.com /2020/07/10/health/what-coronavirus-autopsies-reveal/index.html.

16   Lenny Bernstein, "More evidence emerges on why covid-19 is so much worse than the flu," *Washington Post,* May 21, 2020, https://www.washingtonpost.com /health/more-evidence-emerges-on-why-covid-19-is-so-much-worse-than-the -flu/2020/05/21/e7814588-9ba5-11ea-a2b3-5c3f2d1586df_story.html.

17   Imogen Braddick, "Coronavirus can damage lungs beyond recognition, health expert says," MSN News, June 16, 2020, https://www.msn.com/en-gb/news /newslondon/coronavirus-can-damage-lungs-beyond-recognition-health-expert -says/ar-BB15ysYo.

18   "COVID-19 Had Us All Fooled, But Now We Might Have Finally Found Its Secret," https://www.survivaldan101.com/covid-19-had-us-all-fooled-but-now-we -might-have-finally-found-its-secret/.

19   AS Zubair et al, "Neuropathogenesis and Neurologic Manifestations of the Coronaviruses in the Age of Coronavirus Disease 2019," *JAMA Neurol.* Published online May 29, 2020, doi:10.1001/jamaneurol.2020.2065.

20   Marina Pitofsky, "Illinois reports first known COVID-19-related infant death in US," *The Hill,* March 28, 2020, https://thehill.com/blogs/blog-briefing-room/news /490012-illinois-infant-dies-of-coronavirus.

21   Children's Health Defense Team, "Inflammatory Syndrome Affecting Children: Kawasaki Disease, COVID-19 . . . or Something Else?" Children's Health Defense, May 14, 2020, https://childrenshealthdefense.org/news/inflammatory-syndrome-affecting-children-kawasaki-disease-covid-19-or-something-else/.

22   Mark Blaxill and Amy Becker, "Lessons from the Lockdown—Why Are So Many Fewer Children Dying?" *Health Choice,* June 18, 2020, https://childrens healthdefense.org/news/lessons-from-the-lockdown-why-are-so-many-fewer -children-dying/.

23   "Scientists hail dexamethasone as 'major breakthrough' in treating coronavirus," CNBC, June 16, 2020, https://www.cnbc.com/2020/06/16/steroid-dexamethasone -reduces-deaths-from-severe-covid-19-trial.html.

24   "Researchers Identify 69 Drugs That Could Help Fight Coronavirus," *VOA News,* March 23, 3030, https://www.voanews.com/science-health/coronavirus-outbreak /researchers-identify-69-drugs-could-help-fight-coronavirus.

25 https://www.bmj.com/content/368/bmj.m1086 https://www.bmj.com/content/368/bmj.m1086

26 Ivan Tkachenko, "A Detailed Coronavirus Treatment Plan from Dr. Vladimir Zelenko," *The Internet Protocol*, April 14, 2020, https://internetprotocol.co/hype-news/2020/04/14/a-detailed-coronavirus-treatment-plan-from-dr-zelenko/.

27 Ralph Ellis, "The Lancet Retracts Hydroxychloroquine Study," Web MD, June 4, 2020, https://www.webmd.com/lung/news/20200605/lancet-retracts-hydroxychloroquine-study.

28 Dr. David Brownstein, "85 COVID Patients at The Center for Holistic Medicine: Zero Hospitalizations and No Deaths," LewRockwell.com, April 11,2020, https://www.lewrockwell.com/2020/04/dr-david-brownstein/85-covid-patients-at-the-center-for-holistic-medicine-zero-hospitalizations-and-no-deaths/.

29 G Martinez-Sanchez et al, "Potential Cytoprotective Activity of Ozone Therapy in SARS-CoV-2/COVID-19," *Antioxidants (Basel)* 9, no. 5(May 6, 2020): 389, doi: 10.3390/antiox9050389. DOI: 10.3390/antiox9050389.

30 https://naturalhealth.news/2020-05-18-researchers-claim-100-percent-cure-rate-vs-covid-19-ecuador-intravenous-chlorine-dioxide.html

31 Corky Siemaszko, "End of lockdown, Memorial Day add up to increase in coronavirus cases, experts say," NBC News, June 23, 2020, https://www.nbcnews.com/news/us-news/end-lockdown-memorial-day-add-increase-coronavirus-cases-experts-say-n1231802.

32 Erin Banco, "White House's Own Data Crunchers: Southern Counties About to Get Hit Hard," *Daily Beast*, May 20, 2020, https://www.thedailybeast.com/white-houses-own-data-crunchers-southern-counties-about-to-get-hit-hard.

33 Anne Gearan, William Wan, and Jacqueline Dupree, "As coronavirus rebounds, more patients are being hospitalized and capacity is stretched," *Washington Post*, July 2, 2020, https://www.washingtonpost.com/politics/as-coronavirus-rebounds-more-patients-are-being-hospitalized-thats-a-bad-sign/2020/07/02/62f60720-bc4f-11ea-80b9-40ece9a701dc_story.html.

34 "Swedish minister's 'Russian trolls' fanning 5G fears turn out to be… anti-radiation activists led by local granny," *RT*, April 6, 2020, https://www.rt.com/news/485053-sweden-minister-russia-5g/.

35 Frida Claesson, "En person har avlidit till följd av coronaviruset," SVT Nyheter (in Swedish), March 11, 2020.

36 Mike Stobbe and Nicky Forster, "Little evidence that protests spread coronavirus in U.S.," AP, July 1, 2020, https://www.aol.com/article/news/2020/07/01/little-evidence-that-protests-spread-coronavirus-in-us/24542760/.

37 J Xiao et al, "Nonpharmaceutical Measures for Pandemic Influenza in Nonhealthcare Settings—Personal Protective and Environmental Measures," *Emerging Infectious Diseases* 26, no. 5 (May 2020).

38 Russell Blaylock, MD, "Blaylock: Face Masks Pose Serious Risks To The Healthy," *Technocracy News and Trends*, May 11, 2020, https://technocracy.news/blaylock-face-masks-pose-serious-risks-to-the-healthy/.

39 JH Zhu et al, "Effects of long-duration wearing of N95 respirator and surgical facemask: a pilot study," *J Lung Pulm Resp Res* 4 (2014): 97-100.

40 https://www.youtube.com/watch?v=3STOGvsVCPs&feature=youtu.be

41 Wan Lin, "Student deaths stir controversy over face mask rule in PE classes," *Global Times*, May 5, 2020, https://www.globaltimes.cn/content/1187434.shtml.

42 Andrew J. Campa and Kiera Feldman, "Face masks are now a mandatory L.A. accessory. Can we keep covered up?" *Los Angeles Times*, May 15, 2020, https://www.latimes.com/california/story/2020-05-15/face-coverings-now-a-mandatory-l-a-accessory-can-we-keep-it-covered-up.

43 Denis G. Rancourt, PhD, "Masks Don't Work: A Review of Science Relevant to COVID-19 Social Policy," *River Cities' Reader*, June 11, 2020, https://www.rcreader.com/commentary/masks-dont-work-covid-a-review-of-science-relevant-to-covide-19-social-policy.

44 Zy Marquiez, "3 Studies Reveal How Social Distancing (i.e. Social Isolation) Can Increase Mortality | #SocialDistancing," GreenMedInfo.com, April 3, 2020, https://breakawayindividual.com/2020/04/07/13-studies-reveal-how-social-distancing-i-e-social-isolation-can-increase-mortality-socialdistancing/.

45 Peter Sullivan, "WHO official: Asymptomatic spread of coronavirus 'very rare,'" *The Hill*, June 8, 2020, https://www.msn.com/en-us/news/politics/who-official-asymptomatic-spread-of-coronavirus-very-rare/ar-BB15cBHW.

46 Ethen Kim Lieser, "Study Suggests Spray From Toilet Has Potential to Spread Coronavirus," *The National Interest*, June 18, 2020, https://news.yahoo.com/study-suggests-spray-toilet-potential-220000388.html.

47 Phil Shiver, "Ohio school district plans to surveil students with bluetooth tracking devices to prevent the spread of COVID-19," The Blaze, June 8, 2020, https://www.theblaze.com/news/ohio-school-surveil-students-coronavirus.

48 Beverly Jensen, "During Shutdown 5G Being Installed Covertly in US Schools, Dept of Education Directive," OpEDNews.com, March 22, 2020, https://www.opednews.com/articles/During-Shutdown-5G-Being-I-by-Beverly-Jensen-Absence_Dept-Of-Education-ED-gov_Education_Educational-Facilities-200322-906.html.

49 Katie Magnotta, Steve Van Dinter, and Lauren Schulz, "Verizon 5G Ultra Wideband network live in more NFL stadiums," Verizon.com, September 5, 2019, https://www.verizon.com/about/news/verizon-5g-ultra-wideband-service-live-13-nfl-stadiums.

Chapter 13

1 Walter Hadwen, "The Case Against Vaccination," speech given on January 25, 1896, https://en.wikisource.org/wiki/The_Case_Against_Vaccination.

2 Brendan D. Murphy, "Exposed! 5 Historical Scandals That Prove the Fraud of Vaccinations," Wakeupworld.com, https://wakeup-world.com/2016/10/29/exposed-5-historical-scandals-that-prove-the-fraud-of-medical-vaccination/.

3 R B Pearson, Pasteur: *Plagiarist, Imposter: The Germ Theory Exploded* (A Distant Mirror, 2017) 64.

4 Ethel D Hume, *Bechamp or Pasteur?* (A Distant Mirror, 2017) 295.

5 Ibid, p. 296.

6 Ibid, p. 299.

7 Professor Alfred Russel Wallace. *The Wonderful Century* (Kessenger Publishing LLC, 2006) 296.

8 Kevin Barry, "Did a Vaccine Experiment on U.S. Soldiers Cause the "Spanish

Flu"?" *Free Press*, March 29, 2020, https://freepress.org/article/did-vaccine
-experiment-us-soldiers-cause-%E2%80%9Cspanish-flu%E2%80%9D.

9   Angela Betsaida B. Laguipo, BSN, "Coronavirus has mutated into at least 30
    strains," April 22, 2020, https://www.news-medical.net/news/20200422
    /Coronavirus-has-mutated-into-at-least-30-strains.aspx.

10 Children's Health Defense Team, "COVID-19: The Spearpoint for Rolling Out
    a "New Era" of High-Risk, Genetically Engineered Vaccines," Children's Heath
    Defense, May 7, 2020, https://childrenshealthdefense.org/news/vaccine-safety
    /covid-19-the-spearpoint-for-rolling-out-a-new-era-of-high-risk-genetically
    -engineered-vaccines/./

11 Harry Al-Wassiti, "mRNA therapy: A new form of gene medicine," *Medium*,
    December 10, 2019, https://medium.com/swlh/mrna-therapy-a-new-form-of-gene-
    medicine-5d859dadd1e.

12 Lyn Redwood and the Children's Health Defense Team, "The Dengvaxia Disaster
    Was Twenty Years in the Making—What Will Happen With a Rushed COVID-19
    Vaccine?" Children's Health Defense, April 23, 3030, https://childrenshealthdefense
    .org/news/government-corruption/the-dengvaxia-disaster-was-twenty-years-in-the
    -making-what-will-happen-with-a-rushed-covid-19-vaccine/.

13 Michael S Rosenwald, "Last time the United States rushed a vaccine, it was a mess,"
    Washington Post, May 3, 2020, C7.

14 Robert F. Kennedy, Jr., "New Docs: NIH Owns Half of Moderna Vaccine,"
    Children's Health Defense, July 7, 2020, https://childrenshealthdefense.org/news
    /new-docs-nih-owns-half-of-moderna-vaccine/?itm_term=home.

15 Christopher Rowland and Carolyn Y. Johnson, "A coronavirus vaccine rooted in
    a government partnership is fueling financial rewards for company executives,"
    Washington Post, July 2, 2020, https://www.washingtonpost.com
    /business/2020/07/02/coronavirus-vaccine-moderna-rna/.

16 Marco Cáceres, "Healthy Clinical Trial Subjects Suffer Grade 3 Side Effects to
    Moderna's mRNA COVID-19 Vaccine," *The Vaccine Reaction*, May 24, 2020,
    https://thevaccinereaction.org/2020/05/healthy-clinical-trial-subjects-suffer-grade
    -3-side-effects-to-modernas-mrna-covid-19-vaccine/.

17 Lee Brown, "Moderna coronavirus vaccine tester fainted, had high fever during
    trial," *New York Post*, May 27l 2020, https://nypost.com/2020/05/27/moderna
    -coronavirus-vaccine-tester-fainted-had-high-fever/.

18 Bill Bostock, "6 monkeys given an experimental coronavirus vaccine from Oxford
    did not catch COVID-19 after heavy exposure, raising hopes for a human vaccine,"
    Business Insider, April 28, 2020, https://www.businessinsider.com/monkeys-given
    -new-oxford-vaccine-coronavirus-free-strong-exposure-encouraging-2020-4.

19 https://childrenshealthdefense.org/?s=vaccinated+macaques.

20 Marco Cáceres and Barbara Loe Fisher, "81 Percent of Clinical Trial Volunteers
    Suffer Reactions to CanSino Biologics' COVID-19 Vaccine That Uses
    HEK293 Human Fetal Cell Lines," *The Vaccine Reaction*, July 6, 2020, https://
    thevaccinereaction.org/2020/07/81-percent-of-clinical-trial-volunteers-suffer
    -reactions-to-cansino-biologics-covid-19-vaccine-that-uses-hek293-human-fetal
    -cell-lines/.

21 Laurie McGinley, "FDA to require covid-19 vaccine to prevent disease in 50
    percent of recipients to win approval," *Washington Post*, June 30, 2020, https://
    www.washingtonpost.com/health/2020/06/30/coronavirus-vaccine-approval-fda/.

[22] Barbara Cáceres, "OB/GYN Docs in U.S. Want COVID-19 Vaccines Tested on Pregnant Women," *The Vaccine Reaction*, July 6, 2020, https://thevaccinereaction .org/2020/07/ob-gyn-docs-in-u-s-want-covid-19-vaccines-tested-on-pregnant -women/.

[23] Marco Cáceres, "COVID-19 Vaccine Will Likely Be Given Multiple Times, Perhaps Annually," *The Vaccine Reaction*, July 7, 2020, https://thevaccinereaction .org/2020/06/covid-19-vaccine-will-likely-be-given-multiple-times-perhaps -annually/./

[24] Bruce Kushnick and Scott McCollough, "IRREGULATORS Big WIN: We Freed the States from the FCC," Irregulators.org, March 16, 2020, http://irregulators.org/ irregulators-big-win-we-freed-the-states-from-the-fcc/.

Chapter 14

[1] https://www.starlink.com/.

Appendix A

[1] *Bioimpedance: Phase Angle, Nutritional Status Prognostic Indicator*, https://www .ghtraining.co.uk/perch/resources/bodystat-phase-angle.pdf.

[2] Emilee R. Wilhelm-Leen, MD, "Phase Angle, Frailty and Mortality in Older Adults," *J Gen Intern Med*. 29, no. 1 (January 2014): 147–154. Published online September 4, 2013, doi: 10.1007/s11606-013-2585-z.

[3] "Introduction: Phase Angle," https://www.ebiody.com/phase-angle-in-bioimpedance /?lang=en.

[4] Thomas S. Cowan, MD, *Human Heart, Cosmic Heart* (Hartford, VT: Chelsea Green Publishing, 2016).

Appendix B

[1] "The Hepatitis C Research Project," https://www.biogeometry.ca/biogeometry -hepatitis-c-research.

[2] "Dr. Ibrahim Karim's Hemberg Switzerland Project on Reuters," September 30, 2013, https://www.youtube.com/watch?v=bybKl6VUli4.

Addendum

[1] Jasper Fuk-Woo Chan et al, "Simulation of the clinical and pathological manifestations of Coronavirus Disease 2019 (COVID-19) in golden Syrian hamster model: implications for disease pathogenesis and transmissibility," *Clin Infect Dis*, March 26, 2020, https://pubmed.ncbi.nlm.nih.gov/32215622/.

[2] Linlin Bao et al, "The pathogenicity of SARS-CoV-2 in hACE2 transgenic mice," *Nature* 583, no. 7818 (July 2020): 830-833, https://pubmed.ncbi.nlm.nih.gov /32380511/.

Acknowledgments

The heroes of this story are four people who have done more than any others to shine a light on the truth of the Covid-19 story. We hope we have accurately represented the findings of these warriors: Dr. Andrew Kaufman, Stefan Lanka, PhD, Sayer Ji, and Dr. Kelly Brogan. Without them the key points of this narrative would never have come to light.

For those who have helped along the way, we are grateful. This includes Merinda Teller, who helped us find important references, and Leonard Rosenbaum, for his copy-editing and indexing skills. Much thanks to Gerald Pollack for his insights and help.

To Mary Evans, our agent, much appreciation for guiding us to Skyhorse and for her key contract insights.

To the folks at Skyhorse—Caroline Russomanno, Mark Gompertz, and Tony Lyons—it's been a pleasure working with you.

And finally, to our respective spouses, Lynda Smith and Geoffrey Morell, our love and thanks for your steadfast support and patient listening skills while we worked out the details of *The Contagion Myth*.

Thomas S. Cowan, MD
Sally Fallon Morell

Books by Thomas S. Cowan, MD

The Fourfold Path to Healing with Sally Fallon Morell and Jaimen McMillan, NewTrends Publishing

The Nourishing Traditions Book of Baby & Child Care with Sally Fallon Morell, NewTrends Publishing

Human Heart, Cosmic Heart, Chelsea Green

Vaccines, Autoimmunity and the Changing Nature of Childhood Illness, Chelsea Green

Cancer and the New Biology of Water, Chelsea Green

Books by Sally Fallon Morell

Nourishing Traditions: The Cookbook that Challenges Politically Correct Nutrition and the Diet Dictocrats (with Mary G. Enig, PhD), NewTrends Publishing

Eat Fat Lose Fat (with Mary G. Enig, PhD), Hudson Street Press

The Nourishing Traditions Book of Baby & Child Care (with Thomas S. Cowan, MD), NewTrends Publishing

The Nourishing Traditions Cookbook for Children (with Suzanne Gross), NewTrends Publishing

An American Family in Paris, NewTrends Publishing

Nourishing Broth (with Kaayla Daniel, PhD, CCN), Grand Central

Nourishing Fats: Why We Need Animal Fats for Health and Happiness, Grand Central

Nourishing Diets: How Paleo, Ancestral and Traditional Peoples Really Ate, Grand Central

Index